POSITIVE ECOLOGY

Positive Ecology
Sustainability and the 'Good Life'

GERALD SCHMIDT
Independent Scientist, Austria

Routledge
Taylor & Francis Group

LONDON AND NEW YORK

First published 2005 by Ashgate Publishing

Reissued 2018 by Routledge
2 Park Square, Milton Park, Abingdon, Oxon, OX14 4RN
711 Third Avenue, New York, NY 10017, USA

Routledge is an imprint of the Taylor & Francis Group, an informa business

First issued in paperback 2018

ISBN 13: 978-0-815-34626-5 (hbk)
ISBN 13: 978-1-138-61867-1 (pbk)
ISBN 13: 978-1-351-16368-2 (ebk)

Contents

Preface

"What should I care about your environment?" It was while I was still attending high school and had only just appeared on Austrian national television to argue for the importance of environmentalism (in the early 1990s, when this was an issue even for the children's program) that a schoolmate hit me with this question.

It is not necessarily as absurd a question as it would seem to be. Not only is it common to separate our human lives from the workings of nature, and to wish to define who shall or shall not have a voice in the various issues of "the environment" that confront us. Furthermore, to draw the connections between these concerns and personal life still remains a great challenge. (Drawn by cynical attitudes, one is even led to wonder if the world were not saved already if all professional environmentalists actually practiced what they preached.)

One has to navigate carefully between the eco-pessimism of much environmentalism, and the eco-optimism of anti-environmentalism. I have found it particularly troublesome to explain the middle path I have been looking for: In seeking for synergies between ecological sustainability and what makes up a "good life" for the individual, and development/progress for humanity, every position finds points it has been arguing with. Science, for example, has been working in the direction of increasing transdisciplinarity, environmental organizations do not only act as the modern society's Cassandra, and business is issuing sustainability reports, so there has to be some concern for that somewhere in the private sector. The problem, far from the fact that there are increasing movements towards a "middle ground" where interrelations between different concerns are acknowledged, is that the other contradictory position that would also be included is not usually heard. Balance, particularly in the issue of sustainability where it is always precarious at best, appears not to be a favored condition in these times when you seem to be either "with us" or "against us" in so many instances.

Yet, even if everything is not all well (and, I fear, not likely to "naturally" get better if only we continue to muddle through), hope is far from lost. In fact, aside from the babble (of both the agents of economic globalization and its opponents, for example) that is getting much more attention, science and practical initiative are finding approaches to a "cultural" transformation to sustainability that is working towards the kinds of synergies mentioned above. In a more narrative fashion, descriptions could already be found, e.g. in the Lappé's "Hope's Edge" (2002).

My intention for this work, quite simply, is to provide an overview of the scientific aspects of why we should not only all care about our environment, but could also find a better life in our doing so. There have been arguments and studies (even beyond the rather journalistic ones) that it would and could be the case that such synergies existed, but these scientific arguments have not been collected and brought into relation before. Like the question initially put, it is strange but an expression of human nature (to the extent to which such a creature exists), that "saving the planet" is not enough of a motivation for most people, "living well" is rather important, too. (Not to talk of just making a living – ask a former "child environmentalist" like me now trying to turn into an environmental scientist and make a living with it.) Therefore, a more positive, motivational approach is essential.

Naturally, the very logic suggested here can be said to contain an – obvious (though I am actually indebted to a reviewer for this observation) – bias. Still, in having sought not to find "the solution," nor even all the answers, but only first steps to a perspective based on good science that would fall into the up-and-coming category of what I decided to call a "positive ecology" – that is, an approach to science and practice that would support both ecological sustainability and human quality of life (as the well-recognized LOHAS, "lifestyles of health and sustainability," do, but without just having to entail other, further consumption) – it is my hope that I did so *sine ira et studio* – without fear or favor (without too much of either, anyway).

I, for one, have found these explorations across the range of scientific disciplines worthwhile. More, especially including on more concrete practice-oriented issues which fell short in this academic work, is to come. As a first step to going beyond this study, a website (www.positive-ecology.org) has been established to facilitate communication.

Gerald Schmidt

Acknowledgments

As in any academic work, the author is indebted to all those who have worked on the issues pertaining to the present work, even more so since this piece strived to provide an overview over a breadth of issues and disciplines. Moreover, the continual support of my parents, and the lessons of many people need to be acknowledged. Foremost among the latter are the students and teachers who I had the pleasure to meet during my sojourn at Roncalli High School, Indianapolis, IN and, of course, my host family there. The lessons in social justice, to mention just one concrete example, came at just the right time to influence my thinking on environmentalist issues towards a concurrent reflection of social concerns.

The precursor to the present text was written as dissertation in fulfillment of the requirements for the doctorate in ecology at the University of Vienna, Austria. For having supported and encouraged such a, still strangely unusual, work, I am greatly indebted to Peter Weish and Roland Albert, and to the University of Vienna's Faculty of Natural Sciences in general (which did admittedly resist for a while).

However, it also represents a first culmination of (so far) half a life's pursuit of environmental and social considerations, which began in childhood and is apparently here to stay.

Introduction

Setting out to consider a novel approach to analysis and work towards a transformation to sustainability, it is necessary to clarify its two most basic concerns at once: First of all, reduced to its bare essentials it means nothing more and less than the survival of humanity, which may or may not be threatened by the environmental crisis. The latter argument, especially phrased as a need or desire to save nature/the planet (incidentally, just like when it is framed in the form of a denial that there were any problems, or none that would not be solved by techno-economic progress) tends to misunderstand actual conditions, the relationships of what to save and why. Secondly, and maybe more importantly – at least more immediately meaningful, as will soon become clear – sustainability is about the prospect for future progress and (the conditions for) human well-being. The issue of material-utilitarian relationships between humanity and environment plays a major role in this view, but further interrelations of "nature" and "culture(s)," environment and development, in cultural, psychological, spiritual, etc. relations, are also raised center stage.

So far, there has been quite a lot of talk, much of it more concerned with keeping power relations (in politics, economy, and even academic "turf wars") at the status quo, rather than with the true changes that sustainability will require, let alone with action towards them. The widespread adoption of "sustainable development" as a supposed guideline in business, for example, tended towards using it simply as a "green" synonym for – eternal, now: sustained – economic growth. The growing counter-movement to corporate globalization appears more intent on believing that "another world is possible" as well as on working towards it, but activism against something is much more prominent (certainly, but unfortunately not only, in the mass media's reception), than action towards actual alternatives.

Both movements tend to miss the point of sustainability in an ecological-cultural perspective:

First of all, the dependence of humanity on the ecosphere, which is constituted only as a dynamic, living assemblage of diverse ecosystems and the species which, in interaction with (bio-)physical conditions, make these up in turn. "Nature," in this regard, has its own rules and rhythms of which humanity is a part, and which "are as real as those of aerodynamics. If an aircraft is to fly, it has to satisfy certain principles of thrust and lift. So, too, if an economy is to sustain progress, it must satisfy the basic principles of ecology" (Brown 2001: 77).

Secondly, there is a need to consider the actual integratedness of ecology, or rather the ecosphere, and human culture(s) – in its wide definition as a shorthand for human ways of life and of making a living, spanning material, emotional, intellectual, and even spiritual regards – contained within it. This points to the requirement for a functioning, biodiverse environment as a precondition for human development and well-being. The other way round, it also points to the fact that much of the nature which we now regard highly has developed in coevolution with its human inhabitants (but can only continue to do so if human activities become oriented on making this possible). All of the ecosphere is now in some ways affected by us, but unlike other "natural features" we could decide whether or not we want our influence to be for the better or worse, of greater or less extent.

Thirdly, then, the suggestion of sustainability-in-practice is that an integration of human activity into the structure and function of ecological-evolutionary features and dynamics should be the guiding principle. This will require doing many things very differently from today, not just quantitatively but also qualitatively. However, it can open up the possibility for utilizing synergies to improve both the state and continuing flourishing of non-human nature and the prospects of humanity, and including of the individual person.

Taking its cue from the (rediscovered) "positive psychology" (Seligman & Csikszentmihaly 2000), the approach of "Positive Ecology" is that science should not only focus on problems and destructiveness. Rather, in light of the integratedness of humanity and the biosphere in terms of basic necessities for mere survival of the human species, and of possibilities for improving quality of life by a conscious (re-) integration and coevolution of humanity-in-nature, it should look beyond the necessity of a transformation to sustainability for the future of one's children, to the sense and meaningfulness of such a transformation in the present as well as for the future.

As Alfred Korzybski (1958, II, 4 [1933]: 58) noted, "a map is not the territory it represents, but, if correct, it has a similar structure to the territory, which accounts for its usefulness." So science, in trying to provide a map towards sustainability, cannot be more than a map, and a useful one only if it manages to mirror the structure of reality. Therefore, adequately addressing sustainability as an issue that itself connects facets of life hitherto seen (or at least treated) as basically unrelated, will not work without "creating paths" by bringing together the insights of science, humanities, and cultural resources. Further, it will require crossing the boundaries to the "unscientific" individual and cultural (practical: economic, technological, social) aspects bearing on it, both in theory and in analysis. Finally, without an orientation on also "walking paths" by engaging in activism and action, at the very least by showing examples of the work towards sustainable ways of life that is already under way, the scientific work's value would be lost.

Many building blocks for a "meta-field" of sustainability analysis and action are in existence, both in the form of theoretical considerations, and in terms of empirical, practical examples, but in highly fragmented form. This work represents an attempt at bringing into relation these disparate insights obtained in various fields, in particular ecology (and evolutionary ecology), psychology, and (sociocultural, environmental) anthropology, supplemented by existing suggestions and examples of promising "ways to walk." The aim is to produce an integrated outline of the map for the necessary – and promising – transition to sustainability.

Synthesizing works, even more so in such a case of "post-normal science" (Kay et al. 1999) necessary to address sustainability even just in its theoretical construct, and even more so as a process that needs to be engaging, and engaged in, in praxis, do not have an easy standing vis-à-vis the continuing specialization of disciplines. Yet, without synthesis and an eye toward application, science cannot provide a usable "map," certainly not one that is a meaningful and engaging, as well as an empirically reasonable, reflection of actual conditions and ways forward. It is hoped that this work can in fact profit from its position in the no man's land between theory and practice, natural and social science and humanities, by connecting the pieces of the puzzle in a first, rough outline of ways toward sustainability, as well as by providing extensive notes on their particular sources, so as to make both better known and usable. Moreover, because of both scientific and personal reasons, the main orientation is – at least trying to be – on the relevance to the individual person, for it is persons, not abstract entities, that are both the victims and the perpetrators of environmental problems and social disruption, and who will suffer or profit the consequences of their – collective – actions. Social, institutional, etc. entities commonly appear all too powerfully real and removed from human agency, but still are emergent features of human action and therefore amenable to being changed by social change; and on the other hand, individual decisions about one's own life, and even identity, are getting all the more relevant in the contemporary context of "postmodern" and "globalizing" culture, while and even because sweeping, seemingly hard to influence, changes are occurring.

Sustainability Paradox

In both scientific and popular views, the most fundamental areas of human life can be reduced to the duality of nature and culture (in its inclusive sense of anything that falls into the, supposedly unnatural, human sphere). The environmental crisis, and more recently the call for a transformation to sustainability, have brought their relationship to renewed prominence. They also focused both debate and practical activity on these related concerns: Social, economic, and environmental issues would all need to be jointly addressed; environmentalist worldviews were transferring nature to a more central position; and conferences and publications on sustainability worked hard at achieving some kind of consensus on just what they are about. Actual progress in transforming society, on the other hand, seems even farther off than it did when the Rio Earth Summit (formally the United Nations Conference on Environment and Development, 1992) brought "sustainable development" center stage in the political arena.

In effect, "sustainability" now parallels "nature" and "culture" not only in their fundamental character, and in virtual indefinability, but is further dissolved through its haphazard usage. Furthermore, as an issue that conjoins both nature and culture, and would need to span both hard science and utilitarian economics with ethics, morals and emotion, as well as theoretical considerations with their translation into practical application, sustainability analysis and activism is even more of a challenge than "natural scientific" research or the all-satisfying definition of just what culture (or life, for that matter) is. Sustainability may be a necessary keystone for theory, and even more so a guideline to inform practice, but it is characterized and challenged by the paradoxa surrounding it.

In the scientific community, even a tendency to abolish the term altogether because of the discontent with its vagueness and indiscriminate utilization, and to desist from the discourse because of "[estrangement] from the preponderantly societal and political processes ... shaping [its] agenda" (Kates et. al. 2001: 641) can be found. Most saliently, the concept of sustainable development (or the oxymoron "sustainable growth") which has taken center stage in political discussions, was rather exclusively shaped by economic concerns and theories rather than ecology, even though that discipline is universally acknowledged as of the utmost relevance. Moreover, that sustainability would (have to) represent a transdisciplinary concern has regularly been voiced, but in practice it is a veritable minefield duly avoided, with the natural science-oriented perspectives tending to misunderstand the social side of affairs, the socially oriented views (whether scientific or activist) underestimating the importance of ecological realities, and the sciences generally reluctant to "market" their findings and translate them into practice.

In public discourse, the situation is even more confounded, not least because of the simplified and often ideologically distorted reception of scientific findings, as well as uncertainties and conflicting interpretations of data that just are a feature of critical science. Hawken, Lovins and Lovins (1999: 309) capture the situation nicely, stating that "(t)he episodic nature of the news [on environmental problems], and [their] compartmentalization ... inhibit devising solutions. Environmentalists appear like Cassandra, business looks like Pandora, apologists sound like Dr. Pangloss, and the public feels paralyzed."

A recent case in point is Bjorn Lomborg's (2001) "The Skeptical Environmentalist" which basically contained the twofold message that everything is not as bad as commonly portrayed in "the [environmentalist] Litany" (though not quite well either) and that, therefore, the continuation of current developmental paths is not only possible but even necessary to have things get better in the future. The first part of the message "offers a detailed and well-developed antidote to environmental doom-mongering" (Grubb 2001: 1285) that many members of the environmentalist movement would indeed be well advised to heed. The conclusion reached, however, speaks of "a stunning lack of attention to cause and effect" of environmental legislation and policy, and improvements resultant thereof (ibid.), which is particularly telling in that the failure of environmentalist "doomsday predictions" to come about is regularly hailed as a success of the market, but more likely to have come about because of environmentalist concerns and legislation implemented because there dire predictions were voiced. Even worse, the conclusion exhibits a fundamental ignorance of ecology and its implications, at the very least of the impossibility that "development" in its current form continued indefinitely (cp. Rees 2002, Vogelsang 2002), and of the potential not only for ecosystemic shifts, but also of alternative developmental paths working with ecology to prevent likely problems before they arise.

The danger, as Grubb (p. 1286) so aptly points out, lies with the conclusion that some media did indeed condense out of this publication (no wonder since it fits squarely with standard environmental economics, and more generally with that which many people quite simply want to hear): "Many (though not all) aspects of the environment are getting better ... Therefore, environmentalists are stupid ... And technologies will solve any outstanding problems, so we don't need policy," let alone social or economic changes.

This process reflects one of the main aspects of public discourse on environmental issues (and, by extension, on sustainability): the production of a sense of "well-informed futility" in the public (Wiebe 1973). According to this concept, the individual's sense of personal concern and agency in environmental issues is "narcotized" by the mass media's "onslaught of difficult-to-process information" (Shanahan and McComas 1999: 11; cp. Opotow & Weiss 2000, Kaplan 2000). Furthermore, news of environmental problems may eventually even be tuned out of conscious attention in a mechanism typical of human psychology. This denial and psychological repression of problems that appear not to be imminent, and to be adamant to personal influence, is apparently on the rise even because of environmentalists' constant appeals to sacrificial altruism as the basis for environmentally responsible behavior (cf. Kaplan 2000), which seems to argue that the comforts of modern life would have to be given up in exchange for a meager, bleak existence. Current concerns about political and economic security and corporate globalization have further removed "environmental" concerns from public attention (while their interrelation, and therefore relevance, fails to be recognized). Simultaneously, an undercurrent of apprehension regarding social and environmental issues continues to be a strong issue, and support for the alternative ideas of sustainability, at least in theory, would be high.

Surveys underpin these observations with statistical data: The NEETF/Roper Report Cards (NEETF 1999) on environmental awareness in the USA still do not even ask about sustainability. However, the included question on whether or not environment and economy could go hand in hand, which may be taken to be indicative of the nominal attitude towards sustainability, receives a high count of support. Ray and Anderson (2000: 157ff. drawing on an EPA/PCSD study from 1998-99) report similar findings on high public, voiced support of the ideas of sustainability, and mention that the term itself is considered too technical.

A comparable representative survey from Germany (Kuckartz 2000) shows that sustainability is actually – probably not so much in opposition to the concept's widespread academic and all but random public usage, as because of it – a little known concept in this country. However, as in the American studies, its major components such as the call for intra- and intergenerational equity, the latter being the well known call for sustainable development as development to fulfill the needs of the present generation without diminishing the prospects of future generations to fulfill theirs (of the Brundtland Report, WCED 1987) do receive uniformly high support.

Exacerbating the problem of limited knowledge about, let alone implementation of, sustainability even more is the continued focus on activities that are narratively relevant in environmentalist terms (cf. Shanahan and McComas 1999), e.g. activism, recycling, coupled with the unchanged continuance of environmentally more relevant actions, e.g. the rise in sales of SUVs which are perceived only as a (positively connoted) means of transportation.

The statistics on environmental values also show that many people do indeed realize that another, more sustainable, way of life would be necessary, and would probably even be willing to accept change. E. O. Wilson argues that environmentalism "is not yet a general worldview ... compelling enough to distract many people away from the primal diversions of sport, politics, religion, and private wealth" (2002: 40). If it were a worldview that needed to distract from "primal diversions," the outlook would be bleak. This is probably not, in fact, what sustainability needed to do, but what it needed to work with. The problem remains that what you actually do if you want to live sustainably is not yet really known, or as far as it is, not made known in the context of promising appeals to change in the direction of living well or living better than hitherto. Rather, the communication of alternatives, were it exists at all, still follows the course of "classical" environmentalism in seemingly arguing only for the need to give up such ("primal") diversions and pleasures.

In the absence of a strong – let alone positive – vision that explains not only why we should be concerned about, but also how we would be able to make a living, and to profit in other, wider ways from sustainable patterns of living, it is only too easy to continue with business-as-usual. Psychological, social and economic pressure to conform is enormous, particularly when confronted with such a host of seemingly unrelated, more immediate problems, and with the message (inherent in environmentalist catastrophism) that the current situation were better than anything that could come after it. On the other hand, the example of (initially) only a few people, let alone practical changes that can be enacted and followed not because they prove environmentalist concern, but because they constitute an individually felt advantage and progress (and, maybe soon, a new usual way of doing things), can change a lot.

The "Marketing Mistake" of Ecological Sciences

Even while science has always been of influence in the modern world, where it would be most needed contemporarily – in communicating wherefore and how to guide society towards sustainability – and even though "the limited actions ... we already have are very largely driven by scientific data ..., science is not particularly good at selling itself" (Clarke 2002a: 814). Particularly where diverse aims and needs, and scales of time and space (not to forget aspects of social organization) are effective – even though it may analyze these very factors – science has long been all but totally disengaged in translating its findings into recommendations usable and meaningful in daily life. As Rhoades and Harlan (1999: 278) state for ethnobiology/-ecology, they are "plagued by ... a cowardly disdain for application in solving real world problems."

Ecology, too, though recognized as fundamental to the issue, is conspicuously silent when it comes to engaging in the political and social debate on, and oftentimes even practical (local, educational, etc.) work towards sustainability, focusing on more traditional research and activities such as those in conservation biology and analysis after the fact, instead. This retreat of scientific ecology is somewhat understandable considering how "ecology" has been co-opted as a convenient term for anything concerned with relationships between living beings (or even just entities of some kind, e.g. language) and their environment, or simply for environmentalist issues. The latter "popular ecology," oftentimes looks upon science rather dismissively, focusing on cognitive solutions, i.e. the change of our common world view to one that is supposed to be more positively disposed towards the environment and therefore resulting in less destruction (which overestimates the relation between nominal attitude and actual behavior). But this makes it all the more important to engage in the debate.

From a monist perspective of humans-in-environment, as opposed to the "Cartesian" dualism which separates humanity and nature into distinct spheres which only slightly overlap, even "cultural diversity and the loss thereof, with its frequent corollary of loss of traditional ecological (and other) knowledge and practices, can be seen as an integral part of the overall ecological processes affecting biodiversity on Earth." (Maffi 2001: 11). Similarly, Tim Ingold (2000: 60) argues that in such a perspective "there can ... be no radical break between social and ecological relations: rather, the former constitute a subset of the latter."

Scientific ecology alone can, first and foremost, only explain why a transformation to sustainability is necessary – which it does, but is not very successful in communicating. But purely Cartesian empiric science could not provide a quantitative, nomothetical guide to survival (cp. Harries-Jones 1992: 162f.). It is, however, further needed for and capable of informing this discourse with "an imaginative understanding of patterns of survival throughout evolution, and through human history as part of evolution" (ibid.). Even where science alone would seem to be at a loss, at justifying ethical and moral aspects, evolutionary history can be a valuable input (as in informing the biospheric perspective discussed in Chapter 1), and so a transdisciplinary approach linking sciences, humanities, and public concerns and application is promising indeed.

On the other hand, in spite of the view oftentimes found in what may be called "popular ecology," according to which science and its practical application in technology were somehow opposed to sustainability (because of their roots in attempts to control nature), the current crises are "based not on an excess, but on a poverty of reasoning 'as an active process [of] contemplating the consequences of practice a priori'" (Watts and Peet 1996: 261). This is apparent in how uncritically many supposed solutions to environmental problems are presented as the (singular, simple) solution, without taking empirical findings, other relevant fields (whether theoretical or practical), or the wider connections to other issues into account. Another, related, failure of reasoning is that of not taking the ecological embeddedness of human affairs as the real foundation of all (human) life it is, for example in the economic theory mainly shaping policy, which is far removed from ecological (and anthropological – cultural, psychological) reality. The dilemma this currently puts us in was succinctly pointed out by McKibben (quoted in Hawken, Lovins and Lovins 1999: 314) who stated that, "The laws of Congress and the laws of physics have grown increasingly divergent, and the laws of physics are not likely to yield." Yet, policy and economic practice mainly continue as if the latter could somehow be the case, and even (radical) alternative social movements and social sciences tend to forget that the dependence of humanity on the ecosphere (as inseparable part of it) is not an imagined but an inherent condition (i.e., global equity cannot be achieved at the standard of living of "Northern" countries with current technologies and following what is currently termed progress).

This in itself were an issue where ecology needed to argue more exigently, as it turns the true relations in the world upside down. In fact, all of culture (social life, economics, etc.) is a subset of ecology in terms of being founded in this basis, albeit not determined. Therefore, those principles of ecology would have to form the foundation for "sustainable cultures" (as ways of life). However, in spite of the wide recognition of this necessity, it is usually neither argued for strongly enough, nor actually used in suggesting what a sustainable economy or lifestyle would be, especially not in such a way as to be acceptable, let alone attractive (with the exception of ecological economics, which has been moving in the direction of starting out with first, ecological, principles). This reluctance, though understandable, stands in stark contrast to the (perceived) importance of economic science, which is always turned to for advice not so much because of its successes as because there is (so far) no one else to ask (Wilson 1998). (It is interesting and important to note that the danger of politicizing science – which does exist in such calls and does represent a problem – is hardly ever leveled at economics, the science that is most engaged and political.)

In effect, failing to effectively communicate the necessity, possible advantages, and ways toward a shift to sustainability, only vocal support for abstract general principles is quite large (as seen above), but motivation for and actual change are all but elusive – not surprisingly, since it is usually not even realized in the wider public that this is not just one more modern fad in a long line of issues centered around "saving the planet." Rather, it is the continued existence of humanity, an improving quality of life in terms of individual well-being in its intimate interrelation with what is coming to be known as ecosystem health, and the potential for humanity's ongoing development (whether economic, technological or spiritual), that sustainability is very muc about. In spite (or maybe even because) of decades of environmentalist concern, it was only with TIME magazine's feature article on the Johannesburg Summit (World Summit on Sustainable Development; Kluger and Dorfman, 2002) that a popular magazine broke with the common rhetoric and finally stated explicitly that the quest for "saving the earth" in

· fact means the attempt to save the earth as we like and in fact need it, and therefore to save us.

"Positive Ecology"

The relationships between human survival and well-being and "healthy" ecosystems, in contrast to their wide lack in public discourse, do find consideration in some promising initiatives toward sustainability utilizing a perspective of positive relations and synergies, and local and global grassroots movements. These simply fail to get as much attention as political high-level decisions and the bad news that are proverbial good news to the media. "Cultural Creatives" (Ray and Anderson 2000) combining making a living with care for the environment and society, together with (more personal activity- than activism-oriented) elements of the anti-globalization movement (which in this case is misrepresented by this term: it is directed against corporate power and making of its own rules, but all for a globalization from below, in terms of human solidarity and diversity, cp. Kingsnorth 2003) appear to present a major movement. However, they are not organized in any easily recognizable way, if at all, and even difficult to classify, especially in conventional categories (cf. Hawken, Lovins and Lovins 1999, Ray and Anderson 2000). Moreover, reports of these movements, and even accounts of the positive relations of humanity and nature in general, are to be found virtually only in journalistic reports or in fragmentary form in specialized publications, but not easily available, certainly not in a comprehensive form connecting scientific backgrounds and findings, human concerns, and suggestions for the future.

There also exists some, limited, recognition that a change in lifestyles and forms of development (necessary conditions for a transformation to sustainability) may not imply a bleak, austere existence, but may in fact represent a way towards improved personal and global conditions. The United Nations' "Global Environment Outlook 3," for example, mentions that "a realization that changing consumption patterns does not have to curtail or prejudice quality of life, and can in fact do the opposite, must be brought home to the people concerned. There is sufficient evidence that this is the case but no coordinated effort to get the message across has yet been undertaken" (UNEP 2002: 403; the report itself does not, however, state such evidence or even just give sources for this view).

Even in terms of "sustained" economic competitiveness, such a transformation can confer distinct advantages in various regards – straightforward economic performance as well as worker and customer satisfaction, for example – and in various sectors (cp. Hawken, Lovins and Lovins 1999). There are probable (and even necessary) losers as well as winners, however. After all, a transformation to sustainability will have to entail a transition from a simple growth paradigm to an orientation on (more widely defined, not just measured in terms of profit,) progress. Hence, it will be necessary to make it less easy, certainly less accepted, to make a quick profit by exploitation of natural resources (or people, at that). For example, a fishing industry that continues to consider fleets and size of nets the limiting factor when it is the fish already will at some point need to change considerably or fall by the wayside (with the problem in this case being that the alternative that capitalism immediately discovers, fish aquaculture, is itself problematic in both ecological and social aspects; cooperation for conservation and sustainable extraction would be better in both regards, but is not easily achieved). With some creativity and long-term strategy, and preferably if the transformation to a sustainability-oriented economy is started sooner rather than latter, however, the prospects for innovative entrepreneurship may be improved even as sustainability requires an abatement of consumerism. There are indications of a shift from a more conventional extractive economy to one based on the additional values (such as recreational use and long-term, sustainable forestry) provided by forest conservation even in the American Northwest, for example.

A development within the sciences resonating well with initiatives towards sustainability is the above-mentioned perspective of positive psychology, which was a major inspiration for the present research topic (and the obvious source of the term positive ecology). This re-emergent focus, as argued by Seligman and Csikszentmihaly (2000), is meant to provide "a science of subjective positive experience, of positive individual traits, and of positive institutions [that] promises to improve the quality of life and also to prevent the various pathologies that arise when life is barren and meaningless."

Their position on the role the social sciences should play in "(articulating) a vision of the good life that is empirically sound while being understandable and attractive" (ibid.) translates directly into what a positive ecology integrating ecological, evolutionary and anthropological approaches and findings should do for sustainability – and thus, for us. It must be noted, however, that this represents a critical scientific approach with a re-focused perspective, not anything in the wake of New Age's positive thinking which would argue that everything will automatically get better if only we think it so. As such, it should not be considered the proposal of a Panglossian utopia, as if it could end all environmental problems and social inequities at once. However, in addition to the conventional analysis of the paradox situation of the complex challenges surrounding "sustainability," e.g. how environmental destruction is not unequivocally proceeding as environmentalist doom-mongering would have it, nor showing signs of abating anytime soon if current trends continue, even if it would run up against limits, and certainly an unwanted side-effect of having to make a living, "positive ecology" is indeed intended to provide an additional positive, motivating vision for action towards sustainability, and stronger (evolutionary and human) ecological and cultural anthropological foundations to the limited theoretical foundations and empirical examples of ways toward sustainability we have to date.

In spite of promising initiatives, scientific, transdisciplinary developments, and valuable ecological and anthropological findings, an analysis setting out from the perspective of a positive ecology, presenting the integratedness between ecology and culture, nature and human needs and well-being, in a consistent and synthetic manner and developing suggestions for pathways towards a further integration of these two major aspects of human life oriented towards sustainability, has not yet been undertaken, in spite of its potential to show both sense and meaningfulness of this concept, and therefore – hopefully – to present stronger arguments and motivation.

The necessity of drawing on this potential can itself best be explained by an outlook derived from a combination of evolutionary-ecological theory and anthropological findings: Sustainability, in the longer term, is an idea rather like freedom or equal rights, but one that has to be conceived of "not [as] a fixed ideal, but an evolutionary process of improving the management of systems, through improved understanding and knowledge. Analogous to Darwin's species evolution, the process is non-deterministic with the end point not known in advance" (Cary 1998: 12). What can be said "in advance" is that, in order to be such an evolutionary process, development to sustainability has to start out from the ecological foundations of life, with a focus on resilience (the ability of ecosystems to continue functioning even under unexpected stress), and adaptive development (the analogous ability of society to cope with stresses), and their interaction, which issues require redundancy and low dependence of ecosystem function on human input (cp. Marten 2001).

Bearing human psychology – particularly the tendency for denial of threats that are not considered immediate, and seen as outside of one's personal effectiveness – in mind, and in order to provide guidance in the process of sustainability, not only should the (rational) sense of sustainability be formulated and understood (and, best, shown by example), but its meaningfulness in and as life practice should be increased, as well. This aspect of a necessary connection between reasonable and emotionally powerful approaches comes into clear outline when considering how "all traditional societies that have succeeded in managing resources well, over time, have done it in part through religious or ritual representation of resource management. The key point is not religion per se, but the use of emotionally powerful cultural symbols to sell particular moral codes and management systems." (Anderson 1996: 166). After all, the human is a cultural species, so, as Rappaport (1984: 340) reminds us, "it is neither objectively identified organic well-being nor ecosystemic integrity but culturally constructed value and meaning that ... human social systems tend, in the first instance, to preserve."

These theoretical concerns have so far not been brought into a useful synthesis. So, bits and pieces of their connections among themselves and with practical application, in theory and in actual example, are in existence. Even just to come by them, however, is oftentimes a matter of sheer luck. Yet, taking an anthropological perspective with human needs as the starting point and proceeding to cultural aspects, and integrating it with the essential foundations provided by an ecological-evolutionary outlook promises to be a way to consider sustainability from a perspective that is "empirically sound while being understandable and attractive" (Seligman and Csikszentmihaly 2000), sensitive to alternative conceptualizations of progress (e.g. different cultures' varying support for particular aims in life), as well as individual definitions of quality of life, and to the fact that human well-being and satisfaction – not to talk of mere human existence as such, of course – are intimately bound up with biospheric functioning and diversity. Furthermore, just as (basic) needs are universal, so an understanding and practice of sustainability as positive and progressive, in ecological and cultural terms, and in the orientation towards people (or even meaning to the individual person), addresses both "North" and "South," forging connections rather than arguing over single "root" causes or solutions.

PART 1
Contexts and Approaches

A Plurality of Perspectives

In trying to set the stage for the play of human existence, basic assumptions people hold about the world immediately come into play. Such world views are usually taken from traditional sources and learned during childhood. Even though they can include time scales quite outside of daily human experience, e.g. scientific geological or "deep" time, or the millions of years in Hindu and Buddhist cosmology, life is generally spent acting on time frames, scales of space, and levels of complexity more immediately meaningful to us.

Furthermore, the stage on which human life unfolds, which is the Earth and most importantly its biosphere, is commonly considered a mere backdrop to humanity – somewhat important, but less and less defining of human existence, not least because it were rather static. As it turns out in light of evolutionary-ecological findings, however, far from being the static stage the term "nature" commonly evokes, it is a dynamic field of mutual interaction of organisms among themselves, as well as of and with the physical elements of the Earth system. Many of these ecological processes can occur randomly on one level (e.g. mutations in the genome), but constitute non-random patterns through their interaction over time (e.g. adaptation and species formation).

The human species turns out to be a rather recent and very unusual addition: Instead of being physically adapted to the habitats it occupies, its – that is: our – strategy is to use cognition to actively devise cultures (understandings of the world, strategies, technologies) that facilitate survival (cp. Moran 2000), instead. These also include a host of phenomena quite certainly unrelated to immediate necessities for survival and even reproductive success. Still, as beings that have evolved in this world, mainly faced with problems at specific scales of time, space, and complexity, we human beings are attuned to thinking in certain scales, and amenable to react to certain elements of nature with delight or fright (cf. Wilson's "Biophilia," 1984, for example).

The changes in sheer numbers of human beings and in technological capability are now an influence on evolutionary-ecological processes and even on global biogeochemical cycles that would have been – and sometimes apparently still are – considered natural immutable conditions rather than processes prone to change and interdependent. This shift in relations presents humanity with problems at scales that used to have little meaning. Therefore, evolutionary and ecological perspectives, and education about them, play a major role, but so do culturally salient motives and motivations. We can, for example, learn something about ways of life that represent a necessary coevolution of culture and environment from traditional/indigenous peoples. Nevertheless "there is no pure, perfect or easy solution waiting to be found stored in the non-polluted minds of shamans or retained by all-knowing Third World peasant agro-ecologists" (Watts & Peet 1996: 261f.).

Thus, basic assumptions about the world such as the abstract-scientific view on humanity (e.g. general human needs) but also particular cultural perspectives, as well as scientific perspectives such as those that Daniel Botkin described as the "new ecology" (1990) or "new view of nature" (2000), need to be considered, particularly with regards to their potential compatibility. Also, in trying to address the issue of motivation for a transformation to sustainability, a "personal" perspective oriented on the implications of the more abstract perspectives to individual persons in their own (psychological) and social-cultural diversity should be considered.

Similarly, a certain width is necessary in developing approaches to sustainability. This is well recognized in the perspective arguing that there were "three pillars" of sustainability, i.e. the ecological, social, and economic spheres, that had to be addressed. Actually, this particular approach usually is neither "radical" nor wide enough: radical because it is necessary to start "at the roots," which are primarily the ecological and the concrete human aspects (which tend to take second stage to economic considerations), and wider because changes in economics and technology as well as consumption and other normal, daily behavior – management of natural resources as well as of ourselves – will be necessary to truly achieve a transformation to sustainability.

Chapter 1

The Biospheric Perspective

The "biospheric biography" as we now know it presents itself as a multifarious network of overarching patterns rather than the linear progressive sequence it was earlier considered as having been (cp. Botkin 1990, Gould 1989).

A fundamental pattern consists in the development of new solutions to the problem of "making a living," e.g. photosynthesis, the eukaryotic cell, multicellularity, evolutionary "arms races", and novel symbioses and co-evolution, to name the most important ones. Concurrently, side effects of unprecedented scales were oftentimes produced, most notably from photosynthesis which produced oxygen, thus changing the atmosphere. For most organisms then extant, this highly reactive substance was poisonous and therefore caused a mass extinction. At the same time, it opened up the possibility of aerobic respiration, without which there would not have been any organisms with aerobic respiration, including us. Moreover, there would not be an ozone layer shielding the biosphere from UV rays, nor would harmful substances be removed from the atmosphere by oxidation (Alexander, Schneider & Lagerquist 1997). (Interestingly, even such prehistoric atmospheric changes are used to argue that contemporary global change were just natural, most recently in Michael Crichton's novel "State of Fear," published in 2004, failing to consider the aspect of accompanying extinctions and the problems of adaptation that even we would have with it.)

These developments, just like mass extinctions, were followed by the radiation into new forms of life (another common pattern). The most special of them was, arguably, the Cambrian explosion during which all the body plans now in existence, and several more, seem to have (first) appeared (cf. Gould 1989). Furthermore, yet another pattern of diversifications following mass extinctions is established by the ongoing increase in diversity of species (until the next extinction event), but based on surviving body plans rather than producing fundamentally new ones (ibid.). Research by Kirchner and Weil (2000a: 1308), supports this view of the origination of new species being "primarily controlled by the creation of new ecological niches, and new evolutionary pathways for reaching them, by diversification events themselves," rather than by "the emptying of ecological niches through extinctions."

In sum, patterns can – in hindsight – be discerned, but not as a predictable progress, and certainly not as progress towards humanity, let alone one that implied that the human species could never again go extinct. Mark Twain's statement that, if all of evolution was merely a preparation for the human species, then the Eiffel Tower (at his time the world's tallest structure) had obviously been built for the gilding on its pinnacle (quoted in Gould 1989: 45, 2000: 95), is perfectly compromising of this view in its forceful irony.

One of the most fascinating modern insights is that the biosphere's development thus was neither only an adjustment to the physical conditions present on Earth, nor only an adaptation of organisms to each other in evolutionary arms races, but also a process wherein species developed their "ways of living" in co-evolution, symbiosis, and niche construction, and where organic life even influenced the inorganic, physical conditions on this planet – where climate and life have coevolved (Schneider & Londer 1984). And so, as Sagan and Margulis (1993: 350f.) note, "the most important of [the reasons for saving the environment], and least often mentioned, may be the relationship of certain lush regions of the earth and the present biogeochemical regime – not just global climate, but global chemistry – that supports human beings."

More generally, this intricate and integral pattern of interaction also implies that "the nature of nature" is constant change, so that simplistically static, non-evolutionary attempts at planetary management, together with similar utopian views (whether of a cornucopian economy or an unchanging biosphere) are destined to fail. Furthermore, it draws attention to the fact that even the most static-seeming aspects of the Earth system are actually affected by, and in some cases maybe even dependent on, the integrity of biodiverse, natural ecosystems.

In the biospheric perspective, then, the appearance of humanity has so far barely been an episode, though one that is currently in the process of rivaling (other) natural forces in shaping biogeochemical cycles, and catastrophic impact events in causing impoverishment of the biosphere. In the course of deep time, even these impacts would be just an episode – humanity will not manage to destroy the planet entirely, nor even all forms of life. At this point, however, it is time to introduce the human perspective, from which we have to ask ourselves if we, as beings priding ourselves on our cognitive and moral capacities, really want to be remembered as the cause of the sixth great extinction event – if there are people that can remember, then. After all, it is only with humanity that the biosphere, as Wilson (2002b: 132) put it, "began to think."

With cognition, we human beings had come to wonder about our origins, and commonly come to see ourselves as separate from the world, e.g. having been put into it by God. Biology (especially evolutionary theory) is credited with, or reviled for, arguing that nature including human beings as the "third chimpanzee" (Diamond 1992), "simply" evolved without need for a God, and that it were, in consequence (not being the scripture of medieval scholasticism in which to read His intentions), not a source for morals and values (in contrast to how social Darwinism and instances of the naturalistic fallacy that biology is indeed taking a strong stance against, cf. Futuyma 1998, would have it). As a rebuttal of cherished beliefs, which are still holding on not just in religion but in other ideologies and social theories as well, these findings are taken as problematic. Leaving each area – religion and science – its due, however (e.g. following Gould's, 1999, suggestion of seeing both as "non-overlapping magisteria" informing issues in a "fractal" pattern, the one by rationally justifying that which can be empirically treated, the other by arguing in terms of religious values or cultural aspects), the biospheric perspective may serve to help us feel "at home in the universe" or at least on Earth – not as aliens removed from their true relations and aims, but as an element of nature, intimately related and yet in a (precarious) special position.

Chapter 2

Human Perspectives

For all practical measures, deep (geological or evolutionary) time is meaningless to human life: the scientific prediction that the sun will burn out in about four billion years is of no true relevance to the century, at most, of a human lifetime – we cannot build a way of life (certainly not a satisfying one) on it, as little as we can decide how to live by referring to the laws of evolution (this argument, social Darwinism, sometimes still surfaces but is nothing but an instance of the naturalistic fallacy). Yet, in the embeddedness of human life within and dependent on a functioning ecosphere, the relevance of such considerations in terms of human concerns over individual and collective survival and potential – ways of making a living and leading a good life – is extant, and should come to inform our approaches to life and to the world. (Even if we wanted to suggest that humanity and/or all of life would ultimately, because of the above end of the solar "life cycle," have to leave the planet if it is to go on, it still meant that life on Earth first needed to continue to exist as it has to date, maybe even more strongly so.)

In this regard, the biospheric perspective informs the discussion by two essential insights for us humans:

First, it tells us of the existence of complex (impersonal, not conscious, let alone necessarily moral in a human sense) dynamics of constant mutual change in evolution and ecology, which implies that all actions need have uncertain, unforeseeable consequences. This should serve to remind us that caution would be an excellent guide for action in our dealings with the biosphere, particularly as human cognition appears to be a very powerful tool for solving problems of immediate and not-too-complex kinds, but not for the long term challenges of complex evolutionary process in a changeable world. As Leopold put it, in an admonition that falls in line with the second consideration (below) excellently, "to keep every cog and wheel is the first precaution of intelligent tinkering" (1986 [1949]: 190).

Secondly, considering the contemporary diversity of species is the result of millions of years of evolution, we could well see it as a valuable heritage (and "extended family"). Even in just keeping with the idea that a positive ecology should be focused on the human prospect or potential, extinction is tragic. After all, a species lost is a species that nobody will ever get a chance to experience or study anymore. Evolution will replace them with new species, but this is no excuse for it works at time scales quite outside of meaning to us. – As Kirchner & Weil (2000b) and Kirchner (2002) suggest, the time span necessary for the recovery of biodiversity would lie in the multimillion-year range. (Here, too, even if it were not a result of evolution but God's creation made for us, who would we be to destroy it when He made a pact not to destroy His creations again – which is made at the same time at which man is given dominion, interestingly enough.)

Moreover, at some point we would ultimately be unable to continue living without the life-support system that the biosphere – and therefore the species that constitute it – represents, and are quite unable to recreate its diversity. Even if we were, suggesting that we did so would, as Wilson (2002: 130) argues, be akin to suggesting that we did without libraries, art galleries, Shakespeare, the Beatles, etc. because we could do without or recreate something similar.

Most importantly, however, we should finally realize that the dependence of humanity on the biosphere – which includes species we have already lost and are in the process of losing – is not just an ecological or environmentalist, but actually the "real-world view" (Wilson 2002: 28). Furthermore, as the human species evolved in and with the environment, it is only logical to assume that there exist psychological and emotional connections between human beings and nature, as the biophilia hypothesis (Wilson 1984) proposes, in addition to our physiological-biological dependence on the environment.

The latter relationship is particularly telling when considering an example that does not usually come to mind: the large number of microorganisms that live in and on human beings, effectively transforming us into "walking ecosystems" (Dusheck 2002: 57). These are commonly considered unnecessary, potentially dangerous colonizers of the human body (in public opinion about "bugs"), which they can sometimes and to some extent be. Yet, research shows that their occurrence is "not merely the inevitable result of living in a microbe-ridden world. (T)hey are in fact essential for normal development [of most organ systems]" (Dusheck 2002: 57), as well as to the body's functioning, e.g. digestion and immunity (Rawls et al. 2004, Hooper & Gordon 2001). (Those species are highly unlikely to go extinct as long as there are people, but this image does excellently remind of the inextricable link.)

Bringing different findings together, it seems that humanity may be capable of surviving on a biologically impoverished planet (cf. Jenkins 2003) – and that, in fact, we are already seeing a "cognitive adaptation" to more despoiled surroundings as younger generations grow up in such circumstances and, because they are all they know, consider them normal (what Kahn 1999 termed "generational environmental amnesia"). – But neither sustainability, even just as ensuring that the ecosphere remained in a state amenable to human life, nor progress in furthering human well-being will be satisfied in this way.

In spite of our status as a part of nature, human beings commonly define what they and their workings are by setting these apart from any and all that is "nature". This distinction long appeared to be a common or even universal characteristic of human thought, wherefore it was a mainstay of sociocultural anthropology (and it is inherent in the academic divisions between natural science and the humanities), but this "Cartesian dualism" currently "is under serious attack as a framework for understanding the human condition" (Milton 1996: 12; cp. Croll and Parkin 1992, Ingold 1996, 2000).

From an evolutionary perspective, it is quite clear that culture can easily evolve in response to other cultural features rather than in response to the environment (or even the genes of its human bearers/producers). This opens up the possibility that culture need not be adaptively fine-tuned to the environment, and therefore can be as diverse even within the same environment as it is (culture can even be the selecting environment for genes, as even Dawkins 1998: 285, does mention; cf. Durham 1991 for a well-established argument for these processes of coevolution). This does allow for an analysis of culture as a (abstract) system of views and knowledge, but the total separation between humanity and environment it introduces is at least as misleading as earlier conceptions of environmental determinism, and even more so from an ecological perspective (cf. Ingold 2000). – In the extreme of a (misunderstood) radical constructivism in which the world is seen only to exist as a "cognitive construct, we could change it by constructing different truths, different meanings; we could will environmental dangers out of existence through thought alone" (Milton 1996: 54) – but we clearly cannot.

An historical-anthropological perspective best illustrates the varying ways of perceiving the relationship between humanity and "nature" that are currently dominant; it can, somewhat simplistically but nevertheless close to fact, use a distinction between "ecosystem people" and "global economy people," based on Dasmann's (1976a, b) "ecosystem people" and "biosphere people" labels, with the latter changed to use terms with connotations better fitting the respective, contemporary implications of the terms. A similar pair of opposites is found in the common differentiation between "environmentalist" and "cornucopian economist" perspectives employed by Wilson (2002), for example.

"Ecosystem People"

Many historical, but especially traditional societies have generally been living in one or few adjacent ecosystems, dependent upon the resources these provide for their survival. Some trading was commonplace, but only for special products, such as salt because it is a necessity and quite easy to transport, or luxury goods because of their value. These societies' people are what Dasmann (1976a, b) labeled "ecosystem people."

They may have been able to survive simply because their population densities were low enough to stay within the levels of the ecosystems' productivity and resilience, and/or by migrating when conditions turned unfavorable. However, there is also a distinct possibility that "if they understand [or have been able to learn; my addition] the ecological consequences of their actions, [they] might be expected to take care not to destroy it" (Milton 1996: 29).

The actual historical record is, at first sight, ambiguous. Many species have apparently gone extinct due to human influence – but for the most part during the human spread over the globe which is not a condition of "ecosystem people" but rather of migration in which ecological consequences were either not considered or could hardly have been known. Similarly, even some of the great civilizations which developed and had been prospering for some time in specific environments (and therefore fit the ecosystem people label better) appear to have degraded them to such an extent eventually as to cause their own downfall: Rapa Nui, the Easter Island, is one prominent example for this occurrence (cf. Ponting 1991). Another is Mesopotamia's Akkadian empire, for which recent evidence points to abrupt climatic changes and attendant droughts that surpassed the society's adaptive potential as the primal cause of collapse (Weiss & Bradley 2001). The case is similar for the Maya, the downfall of which in the so-called Classic era has been attributed to environmental degradation, especially loss of soil fertility due to agricultural intensification above sustainable levels (as well as other factors; cf. Sabloff 1990). Jared Diamond (2005) provides a series of case studies on the – oftentimes initially environmental/ecological – conditions for societies' collapse (including the consideration of what this forebodes for our future).

Cases such as those of tropical forests and African savannas, on the other hand – the ecosystems in which human beings have been living for the longest time, and therefore the paragon examples in which to expect an "ecosystem people" view to develop – point to the possibility of coexistence and the actuality of coevolution of human cultures and ecosystems. As Primack (1998: 539) comments, "rather than being a threat to the 'pristine' environment of tropical forests, traditional peoples have been an integral part of these forests for thousands of years." For example, it appears that in spite of the ecological problems just mentioned, it had been possible to support the high population densities in the tropical Maya lowlands (of the Classic era) as well as the region's high biodiversity (Gomez-Pompa & Kaus 1992). An analogous situation of tropical forest lands having supported high population densities in some areas while not losing their biodiversity or capability of regeneration increasingly appears to have been extant in areas of the Amazon prior to 1492 (Heckenberger et al. 2003), as well as in other (now "virgin") tropical forest areas (Willis et al. 2004).

Anderson and Nabhan (1991) even argue that "active conservation management rather than passive acceptance of declining resources characterizes the behavior of numerous cultures from the Amazonian rain forests to the North American deserts," some of which have been guiding their actions by an explicitly "sustainable" orientation, e.g. the (Native American) Seneca with their well known trope of considering the effects of actions "unto the seventh generation." A similar argument is, for example, made by Gadgil and Guha (1992) for hunter-gatherer and stable agricultural societies of India, as well. The same could be argued, *cum grano salis*, for other traditional cultures, considering how cultural/linguistic diversity and biodiversity seem to be commonly correlated, as strongly argued by Nietschmann (1992: 3; cf. Posey & Dutfield 1996) in his "Rule of Indigenous Environments": "Where there are indigenous peoples with a homeland there are still biologically-rich environments."

Moreover, it increasingly turns out that many or most of the world's ecosystems which are considered "natural" or even wild, that is unchanged by human influence, were in fact shaped by human modification to some extent (cf. Posey & Dutfield 1996, Posey 2001, Maffi 2001). Apparently, these regions were modified in even less drastic forms than e.g. the Mediterranean region which was widely shaped by deforestation – but even here cultural landscapes commonly considered natural, and indeed rich in biodiversity, do occur.

It certainly is not enough, however, to draw on the myth of primitive ecological wisdom and argue that "indigenous" always implied "sustainable," nor that regeneration occurred and therefore would do so again whatever we did. Culture is always a more complicated issue, its relation on environment dependent on more factors that just (cognitive) culture. So, the perspective of "ecosystem people" described here, even where it does occur, has to be understood as the abstraction from the actualities of life it is: rather than a paradigmatic world view or a fixed system of knowledge which determines behavior, which would make it relatively easy to transfer it to other settings or to transform it into "global" knowledge, it represents traditional ecological knowledge contingent on and transmitted in the wider culture, including cognitive-symbolic constructs, as well as people's practices and interactions. As such, it is better understood as the abstraction of what is actually a kind of performance in which traditional skills and techniques adapted to the ecosystem's management (developed in and from past experience) are being creatively adjusted to the present needs, rather than strictly reproduced (Richards 1993: 62).

Therefore, a simplistic call for the adoption of "ecocentric" world views or the like, as commonly voiced by some currents of the environmental movement, cannot be "the" solution without its translation into actual ways of life, particularly since human population numbers, and (difficult to define) life-improving technology which we wouldn't want to (or should not) give up, prohibits a return to a subsistence lifestyle of the past. Nevertheless, the possibility for coevolution of humanity in and as part of the biosphere may serve as a reason for guarded optimism and a useful – indeed, necessary – inspiration for a new "forwards to nature" with sustainability.

"Biosphere (Global Economy) People"

The form of human social/economic organization that has developed most recently is the global(izing) system of capitalism and industrialization which, in its relation to the environment, is characterized by its drawing on the entire biosphere for resources.

Therefore, "biosphere people" (Dasmann 1976a) living within this global industrial system are largely removed from the vagaries of changes in the ecosystem that forms their immediate environment. One basic tenet, almost dogma, for adherents of the view that the global economy was the only pathway to global happiness is that the failure of doomsday predictions such as the classical one of Malthus (1798) or the modern classic "The Limits to Growth" (Meadows 1972) were proof that there are no limits at all. Even if their existence is acknowledged, "['cornucopian' economists] just don't like to spend a lot of time thinking about it" (Wilson 2002: 28). Indeed, they assume that it would be unnecessary, because market prices will prove effective in regulating their use or substitution, and technological solutions/replacements can and will be found when they prove necessary and cost-effective.

Moreover, the techno-optimism draws on the belief that the relationship between development, i.e. economic growth, and environmental conservation will follow the course of the Environmental Kuznets Curve (EKC). That is, once (economic) development is at an advanced level, people will be willing and able to pay for the environment, and it will be restored (it should be noted that this is only one possible way in which the relationship may work; the EKC leaves the mechanism open).

In fact, it is only too common for bureaucracies to see economic growth and environmental protection as issues that are not mutually exclusive only in the sense that the former will eventually pay for the latter. The EKC relationship between income and environment, however, appears to only hold true for some pollutants, making it a dangerous guide for generalizations (cf. Arrow et al. 1995), particularly if, as argued by de Bruyn and Opschoor (1997) and de Bruyn (1999), economic growth may become re-coupled with environmental deterioration. Furthermore, for the relationship to work in practice would require that the earth is infinite so that everybody could get rich with industrialization or the like. It also presupposes that ecosystems and species can be taken apart and reassembled like machines, or their services substituted for by technological fixes.

The related debate on whether overconsumption or poverty were the root cause of underconservation is similarly besides the point, as both can obviously have that same effect.

Global economy people's arguments fail to understand that not only the renewable resources, including our food, but even the physical conditions we depend on are contingent on biological diversity and ecosystemic functions which cannot be restored easily, if at all (cf. Daily 1997a; Jordan 1988; Uhl 1988; MacMahon & Holl 2001), and certainly are not easily replaced. Therefore, taking a humans first perspective (that is, the argument that we needed to consider people first, e.g. by funneling available monetary resources to immediate development rather than into measures for abating potential climate change) into a temporally wider view inevitably leads to the sustainability perspective that human needs and ecosystemic functioning must be considered in conjunction. In doubt, a stronger focus on conservation of the latter would even be advisable (since there cannot be human life without nature), but present, human, concerns need to be considered as well, of course – the resolution of both challenges lies in considering synergies, not in squaring both issues off against each other (cp. Dasgupta 2001).

The global industrial system's utopian program incidentally shows a fundamental similarity with the planned economy's utopia. After all, as John Gray, Professor of European Thought at the London School of Economics has argued, both "in their cult of reason and efficiency, their ignorance of history and their contempt for the ways of life they consign to poverty or extinction, ... embody the same rationalist hubris and cultural imperialism" (1998: 3). This may be surprising at first, but is understandable in light of their common philosophical-historical background in enlightenment thinking which had been influenced by evolutionist ideology more than by ecological and cultural reality. The problem is not that the enlightenment call for knowledge and rationality needed to be given up in favor of a mystical view, however, but that a rationality unguided by empirical – ecological and anthropological – considerations exhibits a "poverty of reasoning" (Watts and Peet 1996: 261).

Fact of the matter is not only that non-negotiable limits are inherent in our ultimate dependence on the biosphere, but also that they are set in the fuzzy way that follows from the inner workings of ecological systems. Therefore, it is possible for some people, for some time, to exceed them through the appropriation of productive capacities from distant ecosystems, depletion of natural capital, and through the utilization of non-renewable resources. The resilience of ecosystems further misleads by giving the impression of a "benign nature" that no attention need be paid to by making it possible to stress ecosystems for a long time without destroying their current state. Resilience is not infinite, however, and ultimately small additional disturbances may cause rapid shifts to another, degraded, state (Scheffer et al. 2001).

With the global/biosphere economy, as Kay Milton (1996: 30) reminds us, the whole biosphere has effectively been turned into a single ecosystem of sorts. Thus, the argument of infinite substitutability of resources now becomes a gamble with the entire humanity and with Earth's entire biological heritage based on an unreasonable rationalist hubris (Cairns 2000) and a unilinear view of progress failing to even consider alternatives, that we are ill-advised to risk. – Even if direct resource dependency should really be less of an issue for the modern economy, although it probably still is for the satisfaction of the most basic needs (Star Trek-style food replicators are a long way away), the impact of substances it adds to ecological processes (such as from fossil reserves), and the change of (global) ecosystem properties are still ecological dynamics of great effects. The, mainly non-replaceable, ecosystem services contingent on ecological patterns and processes still are the real sources of the living conditions that human life and economy are based on.

Chapter 3

Biospheric Dependence

The Biosphere 2 project in Arizona was the most ambitious project attempting to mimic the Earth's "life-support system" in structure and function at a smaller scale and by technological means, "with the objective of creating and producing biospheres" (Dempster 1991), to date. A fascinating construction of grand design, its conception was deceptively simple: As directed by an interdisciplinary team, put samples of Earth's "seven major biomes: tropical rainforest, tropical savannah, marsh, marine, desert, intensive agriculture, and human habitat" (Dempster 1991) into a structure closed to material and open to energy exchange. Then, seal it off and see how it develops as a "stable, complex, and evolving" system (ibid.) while supporting human "Biospherians."

Eventually, the original concept was abandoned, and even the attempt to turn Biosphere 2 into a respectable laboratory for biosphere science (cp. Vogel 1998) has since failed. The range of problems encountered, which was a (if not the) main factor for Biosphere 2's failure, is nearly as impressive as the grandness of the design:

- a drop in oxygen concentration from 21 to 14 percent
- a corresponding increase in CO_2 levels
- an increase in nitrous oxide concentration
- the extinction of 19 of original 25 vertebrate species
- the extinction of all pollinators
- a population explosion of crazy ants and of vines
- nutrient overload of water
- a large temperature difference between ground and top (tree canopy)
- ongoing fluctuation of carbon dioxide content

(Data from Daily 2000: 236, Cohen and Tilman 1996, Vogel 1998).

Some of these problems may have been prevented had there been further scientific input, but in all likelihood, a viable system supporting human life would not have been developed (Cohen and Tilman 1996). As Cohen and Tilman note, in spite of many doubts and a fair amount of criticism, there is one simple and important conclusion that can be drawn from the Biosphere 2 project: "At present there is no demonstrated alternative to maintaining the viability of Earth... Despite its mysteries and hazards, Earth remains the only known home that can sustain life."

One may add that there are reasons for the opinion that simplified ecosystems can, at least at the small to regional scale they operate on, uphold human life and function better than Biosphere 2 did (cp. Jenkins 2003, for example). – This does not, however, diminish the argument of human dependence on the biosphere considering "eco-cultural" values other than mere human survival, nor even global ecosystemic functioning.

Yet, as Gretchen C. Daily (1997a: xv) points out, our society's dependence on the biosphere is very little appreciated in the public. In fact, the difference between public appreciation and actual importance is nowhere more obvious than in the case of ecosystem services, which is the common label for all the "conditions and processes through which natural ecosystems, and the species that make them up, sustain and fulfill human life" (Daily 1997b). Among these (cf. Daily 1997a, 2000) are:

- protection from UV rays
- moderating effect on climate, globally and locally
- mitigation of floods and droughts
- purification of air and water
- detoxification and decomposition of wastes and translocation of nutrients
- generation and renewal of soil and maintenance of soil fertility
- pollination of crops and natural vegetation
- control of the vast majority of potential agricultural pests
- dispersal of seeds
- production of diverse goods/resources currently utilized, including timber, forage, pharmaceuticals (including traditional medicine, cf. Farnsworth 1988)
- maintenance and development of biological heritage (cf. Wilson 1988), including relatives of cultivated plants regularly needed to improve cultivars (cp. Iltis 1988), potential new pharmaceuticals (cf. Farnsworth 1988), and potential agricultural products or (resources for) industrial products (cf. Plotkin 1988).

What may also be included in the list are elements of biophilia (cf. Kellert & Wilson 1993), such as the provision of emotional experiences, stress reduction, enhancement of cognitive functioning, and intellectual stimulation.

In the wider public, the value of these services is commonly taken for granted, as was legitimated when the human population was so low that its impact could not seriously undermine their provision (which, as the historical perspective shows, most of the time held true only for some hunter-gatherer groups). Nowadays, typically only after having lost ecosystems locally providing such services (Daily 1997b), an understanding may be on the increase but still is not common knowledge. Certainly, it is not yet at the point of being a major consideration either in economic and political theory or in practical decision-making.

Attempts at introducing economic (monetary) measures for ecosystem services' value have been undertaken mainly since 1997, after (and with) the publication of "The value of the world's ecosystem services and natural capital" by Costanza et al (1997). This work took a global perspective and an approach like that of classical GNP accounting to arrive at a figure for the global monetary value of natural biomes which, as the authors argue, "is better than no number at all" (quoted in Garwin & Masood 1998) even if it is certainly only a first and rough estimate. Notwithstanding the various critiques by economists, the paper influenced further research. The difference between the neoclassical economic approach and the ecological economic perspective can best be discerned by comparing the latter with a "classic" text (or rather critique) of the former:

Standard economic arguments are still reminiscent of Colin W. Clark's (1973) example of the maximum profit theoretically achievable from the blue whale when using "business accounting in the service of barbarism" (Wilson 2002: 113). Clark showed that, when a certain (not uncommon) rate of discount were applied, it would appear to be more rational economically, i.e. profitable, to kill off all the remaining blue whales and invest the money in stocks rather than let them recover and harvest them sustainably. It is possible to counter some of the negative image economic theory suffers from this example by applying one of the same criticisms that were leveled at Costanza's paper, i.e. that it does not even take rising marginal value (the increase in value occurring as a commodity gets scarcer) into account (cf. Garwin & Masood 1998; – let alone other, e.g. non-utilitarian valuations). However, a company's (or even a nation's) business accounting does not include the ecosystem services' values (or costs resulting from their loss) even today, as these are externalities to companies (and nations), provided to the wider public rather than to any specific owner. In the case of GDP, the expenditures necessary to compensate for their loss would even be counted as contributing to an increase, rather than figure as a preventable cost, in addition to the rise resulting from profits made in destroying the service.

"Economic Reasons for Conserving Wild Nature" (Balmford et al. 2002), a recent report in the tradition of Costanza's paper, takes a step forward by not comparing "the gross values of the benefits provided by natural biomes, but rather the difference in benefit flows between relatively intact and converted versions of those biomes" (p. 950). The report's findings suggest that, in estimate "the overall benefit:cost ratio of an effective global program for the conservation of remaining wild nature is at least 100:1" (p. 950); but the conversion of "natural" habitats is continuing because these benefits are not counted in usual cost-benefit analysis.

The reason is one that surfaces constantly when sustainable development is considered in the political arena: Conservation will be necessary and of benefit for all in the long run, but more immediate concerns (of development, economics, etc.) are higher on the public agenda. Aldous Huxley, in 1944, already wrote that "we immolate the present to the future in those complex human affairs, where foresight is impossible; but in the relatively simple affairs of nature, where we know quite well what is likely to happen, we immolate the future to the present" (1994 [1944]: 266).

As long as ways to make a living with and possibly by conservation – sustainable livelihoods – are all but known, not much will change. That concern applies to industrialized countries, which should take an exemplary role but largely fail to do so, as well as to developing countries which so far tend to closely follow the (prior, but currently slowly changing) example of industrialized countries. However, there is a widespread recognition (still not influencing actual policy-making, however) that the latter path of global industrialization would require more than one earth to be possible (cf. Wackernagel & Rees 1997), so that (barring resignation) different solutions will be necessary.

Chapter 4

The Paradox of the Human Position

In the current discourse on sustainability, diverse traditional and environmentalist ("ecosystem people") views and the "cornucopian economist" (Wilson 2002) or "biosphere people" perspectives tend to clash, and claim superiority over their respective opposite. The environmentalist approach suggests that worldviews (or paradigms) closer to the view of ecosystem people, which are based on acknowledgment of human dependence on the biosphere, would need to supersede the dominant dualistic paradigm, which sees human beings not only as separate, but because of recent technological prowess even as superordinate to "nature" – e.g. dominant social (human exceptionalist) paradigm vs. new environmental paradigm (Dunlap & Catton 1979), technocentric vs. ecocentric values (O'Riordan 1976), ego-/homocentric vs. ecocentric ethics (Merchant 1992). The hierarchical relationship between humanity and "nature" these contrasting perspectives suggest is particularly telling, for it puts its (especially recently) paradox state – that neither traditional nor modern perspectives are, for the most part, good at conceptualizing – into focus.

The evolutionary-ecological approach to sustainability strongly raises this challenge of the inadequacy of both environmentalist and "global economist" versions of contemporary environmental (mental) maps to the current predicament:

Humanity may not just be engaged in a struggle for survival in a hostile world (that certainly would and could not need protection from man) anymore, but be a major factor of influence on many ecosystems, and even the main force in human-dominated ecosystems. A closer look at evolutionary ecology does, however, disprove the assumption that only human beings were engaged in some such behavior, as ecological niche construction is a common element of these processes (cf. Odling-Smee, Laland, Feldman 2003). Its role in human ecology is, of course, different in that it is (potentially) an issue of conscious reflection, and much stronger because of the application of technology. In synthesis, human activities are still "natural," but that does not mean that they needed to contribute positively to long-term survival/sustainability, and certainly not in such a way as to free ourselves from responsibility for, let alone the consequences of, them.

At the same time, atmospheric phenomena, natural disasters, and the like can and will not be controlled, pathogens do not wait to be eradicated but evolve in response to our misguided efforts to do so (misguided for example when the exaggerated use of antibiotics in the "war on germs" produces more resistant microbes, therefore increasing the danger they pose), and the simplification of ecosystems may endanger the provision of ecosystem services which have always been taken for granted but, as already shown above, are not.

Furthermore, the very question whether or not humanity were able to survive in a biosphere of purely human design, or rather just biologically impoverished because of human activities so that it were not containing wild places (mainly) left to themselves anymore, only reflects the common utilitarian perspective which is eventually inhumane. For neither does this approach value human life in all its aspects, nor any other forms of life, forfeiting the human inheritance from and to other generations (to put it into anthropocentric terms nonetheless). At the same time, it might even produce the conditions for increasing psychopathology (cf. Gullone 2000). – After addressing more utilitarian aspects of the relationship of humanity to the biosphere in Part 3 of this work, it will also present the wider cultural, psychological, and spiritual aspects which make humanity more than just animals, but which a framing of the issue in terms of the predominant – economist – paradigm cannot include, in Part 4.

Ultimately, the question is not one of whether or not there may be human influence, it is what shape it can take so that "both people and planet" profit rather than have to be squared off against each other (which would be a faulty view, anyway). Here, the perspective on indigenous peoples' conservationist tradition and suggestions for the future need to understand that there have always been transformations, even in the "wilderness" of the Amazon, etc. (cp. above). But, as Moran (1996: 549) notes for this exact case, "native systems seem to have been able to transform the environment while preserving some of the features that we all value in tropical rain forests;" finding the adequate balance and orientation between transformation and conservation is the challenge, a challenge all the greater as long as "extremist" trains of thought hold their sway.

The paradoxical state of the relationship between human agency and prowess on the one hand, and utter dependence on ecological functioning, on the other hand, is reflected very well in the juxtaposition of the "Western" conception of the individual as an autonomous island unto him/herself, with the observation that "all but a few eccentrics who subject themselves to 'wilderness training' would quickly perish if deprived of the goods and services provided by their fellows" (Farris 1984: 132, quoted in Atran 2001: 165). Similarly, and more immediately relevant to sustainability, both poverty (even if it occurs while living in more natural/wild environments and knowing the dependence on them well), and affluence (whether disconnected from nature or expounding deep ecological principles) must entail effects/impacts, and can therefore be destructive of the environment. – From this perspective, again, the challenge will be to make sustainable ways of life achievable and attractive.

New, or better still known and valued but re-interpreted, world views subsuming the paradoxical character of the human position within the environment as well as the evolutive "nature of nature" would be a step forward, but it need also be recognized that different perspectives do not immediately translate into different actions. Therefore, a comprehensive approach spanning (the theoretical and practical precedent of) wider and personally engaging "cultures of sustainability," inclusive of individual agency, human nature, and possible social temperance by a budding global citizenship, scientific reason and culturally (e.g. religiously) salient aspects, as well as addressing the potential for (making a) living in ways conservative or restorative rather than destructive of natural capital, is certainly more promising than the mere quarrel over the effectiveness of one or the other of these approaches. Biospheric and human perspectives can and should be utilized, but also gone beyond as their most important contribution would (need to) be to inform personal, individual views and action, which does not come about naturally if support and/or examples are missing.

Chapter 5

"Personal" Perspectives

Abstract, generalized human perspectives and, more recently, the biospheric perspective have been elements of philosophical and scientific discourses for all of their history, but their linkage to the individual person is only achieved by statistical reductionist approaches. The particularistic aspect of how scientific findings could and probably should inform individual lives, together with culturally learned dispositions, religious ethics, etc., however, is largely outside of the usual scope of (objective, theory-oriented) scientific investigation. The same problem – one is tempted to say as a matter of course – applies to any generalizing perspective, whether of population dynamics, human history, or various suggestions of future developments (literary, philosophical, and scientific alike, exhibiting the same tendency toward the utopian or dystopian extremes).

As motivational as utopias can be, they have come to be regarded with skepticism – rightly, considering how inhumane and extremist their proponents tend to become (e.g. arguing that people need to be brought in line by any means, and the greater the aim, the more extreme the means seen as justified). Nevertheless, what approaches can most likely open and ensure the existence of a wide range of possibilities for the individual to pursue in working towards a good life, as well as to contribute to the continuance of a diversity of possible choices, is a major issue. Not least, it introduces a notion of a necessary social justice into the discourse of sustainability, not just as a theoretically good ethics, but (also) more practically because people would need to contribute in order to feel – and indeed be – involved. Of course, the diversity of views of the future and of progress, and the tendency to be both ignorant and in denial of the empirical hierarchy that would (in light of the biospheric dependence) need to give ecological considerations highest importance, do not offer a simple solution to achieving empirical justification and an approach that is socially just. Ecoregion- and community-based projects in collaboration between scientists and local people, however, do offer some hope, as would a more widespread education for sustainability.

Particular in considering sustainability, those individual-oriented perspectives may have been forming the (humanistic) backdrop to some developments, but received rather less or only indirect importance. Notwithstanding this tendency, informing individual behavior will be one of the most important aspects, as it is the countless individual decisions, whether small personal ones of daily life or strategic or political decisions affecting whole companies and nations (which are still made by human beings, not by abstract institutions, even if they do take on dynamics of their own), that ultimately make up a large part of the course of human history – and future.

The confrontation with the global environmental crisis, as noted in the introduction, typically results in feelings of futility and desperation, exacerbated by the way in which the focus of reporting has been on catastrophic events, pollution endangering health, and economic and political decisions apparently outside of control. Issues of the relationship between the environment and human life, considering not only the dependence of the latter on the former, but especially quality of life and the potential for increasing human well-being, just as positive developments currently under way, have largely gone unnoticed. However, the way in which initiatives towards sustainability can improve the chances of people of underdeveloped countries for improving their living conditions, and may enhance the quality of life even of people in industrialized countries, is promising for the project of a positive ecology.

There appears, after all, to be a widespread doubting of humanity's future, even to the point of this generation being unsure whether there will be a future for it or its descendants at all, even though "the hope for a better future is a characteristic of the human makeup that has been found across a wide range of cultures" (Kaplan 2000: 495 quoting Cantril 1966). The realization that this is a result not of uncontrollable external forces but rather of the way much of humanity currently makes a living is depressing. However, there is also the (not yet widely explored) possibility of working towards sustainability in how one lives and makes a living; and the ability to work in a positive direction true to one's beliefs and hopes for the future liberates a creative potential that would be sorely needed, again no matter whether in industrialized or "third-world" countries. Not least, this approach could go beyond the stale debate between environmental catastrophism and cornucopian economists to a re-affirmation of life and personal engagement in the world. Individual human actions will certainly be an element in shaping one's individual life, and as an element of collective human agency and tempered by ecological processes, and therefore as an emergence from complex sources rather than a straightforward result, of course, they will create the future. Nicely enough, coming around to active participation in (one's) life in and of itself holds a potential for increasing individual life satisfaction by improving perceived agency (cp. Chapter 20).

PART 2
Towards A Synthetic Theory

Map and Territory

Sustainability analysis and activism are characterized by their diversity more than by anything else, not least because the concept exists in various forms emphasizing different issues, e.g. ecological, social, and economic ones in the prominent "three-pillar" approach. The call for addressing those different aspects in their interconnectedness is itself a prominent issue, but usually fails to be fulfilled as one of them is brought to prominence, and the others ignored as being either unproblematic or solved in passing – typically, the Cartesian dualism separating culture and nature into worlds of their own (and other instances of such "extremist thinking"), producing "disengagement, ... the separation of human agency and social responsibility from the sphere of our direct involvement with the non-human environment" (Ingold 2000: 76), is all too strong: Where the "natural" side of sustainability is in focus, science and technology are seen as root cause and/or solution, with human behavior seemingly being unimportant; and the other way round, focused on social issues (including politics and economics), the ecological reality these are embedded in is considered unproblematic in comparison.

Alfred Korzybski's admonition that "a map is not the territory it represents," is less known to continue to state that, "if correct, it has a similar structure to the territory, which accounts for its usefulness ... If the structure is not similar, the traveler is led astray" 1933 (1958, II, 4: 58). Therefore, just as reality is one world that can be deconstructed into constituent parts such as "nature" and "culture" only communicatively-cognitively, but can ultimately be understood from the basis of a monist perspective analogous to reality's structure only, so sustainability is both an ecological issue concerning "nature" and its integrity, and at the same time, a social/cultural issue of how to live sustainably. These two sides can be separated when dealing with particular issues, as they have been in the notorious division between the "two cultures" of science and the humanities – as they were described by C.P. Snow (1959 [1998]). The dualist perspective is under critique as detrimental to understanding in general (Wilson 1998), and especially of the human condition (Croll & Parkin 1992, Ingold 1996, 2000), and particularly problematic in working towards sustainability, however.

First and foremost, sustainability has to be about survival. However, understood from the monist perspective in which humanity and nature cannot be separated – and in which, in light of the biospheric perspective described above, humanity without nature cannot exist, – it is not "the planet" the survival of which is at stake, but rather ours. A misunderstood, anthropocentric, monism that only argues epistemologically that humanity and nature were in fact one, and that therefore the one were interchangeable for the other, will not lead us far; an ecocentric perspective arguing that things would turn out well because or if humanity will cause its own extinction is hardly very motivating, either.

Secondly, after all, the focus of sustainability lies on the human prospect, which – as would go without saying – depends on survival, but transcends it: both in more abstract, long-term, and in more personal, short-term perspectives, our prospects for ongoing cultural "evolution" (whether technological, social or, hopefully, in terms of wisdom) as well as for individual well-being are dependent on ecosystemic integrity and interrelated with "nature" and its diversity. Moreover, with anthropological (psychological, cultural) issues such as the commonalities and diversity of outlooks on what constitutes an individual "good life" and a positive development of humanity coming into play, the map for sustainability turns out to be less a roadmap to the one solution, but rather a picture of "lenses and latitudes": Lenses as essential focal points, mainly of ecological-evolutionary requirements which are integral to the map's paper itself, and human needs, psychology, and cultural aspects which are more shifting foci, and – particularly in their interrelation with nature – possess certain latitudes. As environments differ and therefore require different concrete approaches in particular settings, so guiding ideas and human behavior should take on a certain width, e.g. in their interaction with possibilities, individual preferences, social and cultural demands, and the like. Trade-offs are usually seen to be absolutely necessary. Between individual demands or freedom and both social and ecological "well-being," there does need to be a consensus-building (but this is why politics and social conventions exist). As usually conceived, either by forfeiting human development for the sake of conservation, or denigrating the need for conservation for the sake of (economic) progress, however, it is a dangerous perspective. In a longer-term perspective, after all, no progress can be achieved and sustained if it destroys the ecological foundations of human life – but neither can mutually inclusive ways of life be found if they are not paying attention to human needs (nor are even being looked for).

Conflating development (and progress) with economic growth in usual terms (e.g. with indicators not reflecting the state of natural capital), "sustainable development" is no different from "sustainable growth," which "is not possible because all growth must come to an end at some point" (Southwick 1996: 332). With a redefined approach to development, e.g. utilizing indicators of sustainable human welfare, it "has long-term possibilities, depending upon one's expectations, but it also has limits" (ibid.). Understood as a process in which – more widely, and re-defined – progress is integrated with the (ecological-evolutionary) necessities for long-term survival and prosperity of humanity-in-environment (for which there is no substitute), sustainable development maybe is "not an oxymoron, but represents a logical partnership" (Holling 2000). Actually, a not just economic, but more generally creative, imaginative, and even entrepreneurial, developmental element is inherent in the need for sustainability to be an evolutionary process rather than a static harmonious balance between humanity and nature, which would be impossible in a world that is characterized by change(ability). In this respect, even economic growth, – not as a measure of consumption of non-renewable resources and conversion/destruction of biodiverse ecosystems, but as a measure of human(e) development and progress – could potentially go hand-in-hand with sustainable development (cp. Ekins 2000), but there is little reason to define this progress in purely economic terms and thereby decrease its focus.

Going beyond the foundations in ecology and evolutionary theory to the (social) issues of the need for sustainable ways of living and to promising approaches to developing and promoting them, psychology and sociocultural anthropology offer valuable insights. The concept of culture as used in anthropology, in particular, is useful because it does try to recognize and include social, political, economic, and even psychological, cognitive, and spiritual spheres of life in a holistic manner, helping to avoid the reduction of human life to any single one of those aspects. At the same time, informed by practice theory, the anthropological concept of culture further serves to counter essentializing tendencies by pointing out the actual interaction between seemingly independent spheres such as culture or economics, and human individual and collective agency, thus constituting the potential for a re-engagement of the individual person with and in society and environment.

Chapter 6

Evolutionary-Ecological Exigencies

Since survival is the most basic issue of sustainability, and dependent on the continuance of ecological processes, and therefore on "saving the planet as we like it" (Kluger & Dorfman 2002) and indeed need it (as far as that is possible bearing in mind that change is constant), "the only kind of sustainability that makes long-term sense is ecological sustainability" (Southwick 1996: 332). Scientific ecology therefore constitutes the foundation, and the principles derived from it the fundamental guidelines, of sustainability. – Thus, acknowledging that humanity is embedded in and dependent on the ecosphere and, therefore, that we could not survive without a diverse and functioning nature, an orientation on coexistence/coevolution necessarily forms the new paradigm (cp. Thiele 2001).

The salient nadir of current discourses assuming that the ecological would not be the real-world view is the "poverty of reasoning 'as an active process [of] contemplating the consequences of practice a priori'" (Watts and Peet 1996: 261) – including that human dependence on, and inseparability from, nature is truly existential for our material reality, and even (as will be addressed in the course of this work) for psychological well-being and other – seemingly esoteric? – facets of life. It need be noted that interpreting this relationship as environmental determinism would be unfounded, yet not understanding the existence of the relationships as the real-world view would be delusional.

The absolute independence assumed (with the total substitutability of resources) in neoclassical economics, in contrast, is as ideological as an interpretation of the relationships of humanity-in-nature as determinism would be. Similarly, the influence human activities are exerting on the entire globe, even if they turn many ecosystems into human-dominated (eco)systems, does not mean that we could control the entire ecosphere, or that we could substitute technological solutions for its provision of conditions and services. These views, and the support they easily receive, particularly in contrast to calls for sustainability (dismissed as a romanticizing "back to nature" implying that all modern amenities would have to be given up, or just as wrongly as the continuance of current practices with an environmentalist touch), is rather reminiscent of the child wishing to jump off a high place believing that it could fly. Yet, the fundamental principles of ecology, even if they are not quite as well understood, or as easy to define and understand, "are as real as those of aerodynamics. If an aircraft is to fly, it has to satisfy certain principles of thrust and lift. So, too, if an economy is to sustain progress, it must satisfy the basic principles of ecology" (Brown 2001: 77).

At the same time, we do need to realize that science cannot provide a simple, "empirically proven" solution to the complex challenge of achieving (a transformation to) sustainability, but rather only "an imaginative understanding of patterns of survival throughout evolution, and through human history as part of evolution" (Harries-Jones 1992: 162f.).

This relative indefiniteness of sustainability is problematic because – in connection with the liberal-relativist interpretation of views on nature to mean that it need not concern us, as in the examples just mentioned – it tends to be taken as if it meant that this issue, too, may not need to concern us, so that the importance of ecology is once again denigrated. Yet, even from the perspective of economic theory, the non-ecologically founded concept of "weak" sustainability (based on assumed total substitutability of resources) has serious methodological defects, as it would not allow for empirical resolution of its own basic assumption, while the assumed non-substitutability (of natural and man-made capital) of strong sustainability would allow for the identification of exceptions (i.e. forms of capital which are in fact substitutable) (Ekins 2000: 77). Moreover, the ecological-biospheric considerations described above (e.g. the results of the Biosphere II project) provide further, empirically founded reasons to argue for acknowledging ecological principles as fundamental guidelines, and ecological-biospheric patterns and processes as insuperable foundations of life. In a simple and understandable way, these issues become apparent when one compares the theoretical construct of the nature-humanity/culture dualism (as two largely independent spheres) with the view of the Earth from space as the single world we inhabit; notwithstanding the problem that this topology of the globe can also support alienating tendencies, as Ingold (1996) has pointed out.

A further challenge arises from the stipulation that these elementary (ecological) principles should, even because of uncertainties about ecological processes and in spite of the limited support they are as yet receiving, rather be stricter than may be (or appear) necessary, both in order to err on the safe side (in regard to ecological processes) and to account for the probability of error (in regard to human psychology; as suggested by Anderson, 1996). This also fits in with the suggestion that rather than hoping for the best (e.g. that we would be able to find technological replacements for natural capital and ecosystem services once they become necessary), it would be better to prepare for the worst – in this case, to ask less what elements of nature we could do without, and focus more on asking how we could conserve as many as possible.

On the other hand, such fundamental principles can, in addition to being a factor for motivation, eventually provide the necessary guideline to the likely "sustainableness" of particular practices in terms of their compatibility with ecological (and evolutionary and, possibly, cultural) necessities and foci, even if much work needs to be done to actualize this potential.

Leaving the details of what form these principles' translation into practice will take might even prove to be a boon to action towards sustainability. Far from being less valuable than a singular "scientifically proven" sustainable way of life, which would be likely to exhibit quasi-religious and consequently inquisitorial streaks, a strong "lens" of fundamental guidelines that also focuses on exhibiting a certain "latitude" could not only avoid ideological totalitarianism but increase the potential for being adaptive, responsive to local environmental circumstances and cultural conditions, as well as innovative and engaging.

Taking sustainability-as-survival as the primary concern, and expanding it towards the complementary focus of positive ecology as the development of sustainability in such ways as to profit, in a wide sense, from it, further opens the possibility of (exploring the chances of) redefined progress even without having to know all details with absolute scientific proof. – The challenge of translating scientific certainty about basics such as those of ecological sustainability, including the necessity of erring on the side of caution in influencing "nature," into public/political discourse in which economic "science" is a stronger influence, remains. All the while, an ecological reformulation of economics is under way, at least with "ecological economics" as a discipline aiming to make economy and ecosystems compatible in the long term (Guha & Martinez-Alier 1997: 22; advanced by the journal "Ecological Economics," and recently having achieved the status of a discipline with a textbook, that of Daly & Farley 2003). It may have to entail great changes, but addressing them from a background of ecological principles and sooner rather than later would be the best long-term strategy even for the economy. Hawken, Lovins and Lovins (1999) make that point very clear in "Natural Capitalism" (and on www.natcap.org) by arguing that business were still working under the principles of industrial capitalism, e.g. viewing fishing fleets as the limiting factor on the fish catch, when (ecological) principles are already starting to be the much stronger influence, e.g. the lack of fish in the oceans, and making a "next industrial revolution" a prerequisite for the future.

An example concerning the debate on global climate change also shows why the argument of a necessary scientific certainty is sometimes only misleading: The basic argument tends to be that climate change may not be happening, not be caused by human activities (at all or for the largest part), or prove to be (partially) beneficial, so that a reduction in emissions of climate-affecting substances were unnecessary, particularly as it would hurt the economy – and, from the other side of the argument, basically all the opposites of those points. The (mainly economic) perspective that is, surprisingly, all but totally absent from the debate is that of a somewhat longer time frame:

After all, we do know that fossil resources are not unlimited and that alternatives will therefore be necessary (and actually, even before they run out, sinks for carbon may do so, and problems will arise as the peak of production is passed and prices concurrently start to peak). Just as technological developments improve the possibilities for utilizing unconventional fossil resources (e.g. tar sands), the development of alternative, renewable sources of energy is progressing. An "adaptive" development of such guided by ecological-cultural principles could, far from hurting the economy, provide a positive impetus for human progress e.g. in terms of better (easier, with less impact) providing energy in rural areas, constituting less dependence on geopolitical power games (e.g. considering that the vast majority of known oil reserves are located in Middle East countries), with the side-effect of being less of an influence on the global climate.

That the burning of fossil fuels results in an input of carbon into the carbon cycle, in the form of CO_2, that had not been circulating for millions of years should at least be apparent from a very low-key understanding of science, as well. That this represents a disturbance likely to be of some effect is just as easily understandable. – In fact, according to Mackenzie (2003: 192), human-induced emissions of carbon into the atmosphere amount to 12 percent of current total flux, which "is a significant perturbation".

Moreover, those companies that are most advanced in the alternative sector at the time of need (or before that) will have significant advantages. Royal Dutch/Shell, for example, indicates the validity of that point (at least as guide to long-term business strategy, if not much of current practice – which sometimes does make these initiatives appear as mere "greenwash") by having already started considering itself as an energy-service (rather than merely oil) company and accordingly working towards an expansion of business into alternative energy production, and hydrogen storage of energy (cp. Shell Report 2002, Shell International 2001).

Even if ecological sciences are not good at (exact) forecasting in the form of predictions – Peters (1991, quoted in Simmons 1997: 191) rightly states that, "given the complexities inherent in the rather ragged edges of time and space, it is scarcely odd that ecology should not have acquired a good reputation for making accurate predictions," – possible scenarios can still be developed (cf. Clark, J.S. et al. 2001; Kay et al. 1999) and put to use (cp. Rhoades 2001). Oftentimes, the communication of actual scientific consensus, certainties, and uncertainty is as much of an issue as the actual power of ecological forecasting. Particularly in the context of ecological sustainability coming into conflict with issues of livelihood/economics, the work of scientific (or other) institutions at corroborating the importance of the wider and longer-term issues, and of building consensus or surmounting the contradictions is particularly important (cp. Evans 2002: 17).

Embeddedness

Setting out from the fundamental focus on survival, and acknowledging our dependence on the living world, both that human-dominated (e.g. agricultural) and that natural/wild, the major foundation for sustainability lies with ecology: More than just the relation of limited (non-renewable) resources to economic affairs (a common focus of concerns at least since "The Limits of Growth," Meadows 1972, or even since Malthus' 1798 "Essay on the Principle of Population"), the requirement of ecological conservation and environmental protection, expanded to an integration of the workings of humanity into those of the ecosphere, has to be the cornerstone of sustainability.

The reality of limits is probably the best known "ecological" issue, and indeed "the message of ecology is essentially one of the limits – planet earth is finite" (Southwick 1996: 332). In the case of fossil (and mineral) resources commonly thought of, however, the problem is not so much the resources in and of themselves. The base of non-renewable resources is large and to some extent expansible by technological advances in procurement and efficient use. The state of oil reserves, for example, is such that the relation between proved reserves and current rates of consumption has been consistently indicating a depletion within 20–40 years for the last decades: the discovery of additional reserves off-field and in forms (e.g. oil sands) formerly considered not extractable for lack of appropriate technology and economic incentive, but included in more recent estimates, accounts for the expansion of proved reserves in excess of the growth in consumption (cp. UNDP 2001). Nevertheless a peak is estimated to occur relatively soon, if it has not already (cp. Deffeyes 2003). What is most telling is how strong the reaction to such analyses are, continuing to deny their validity even if it is not really likely that oil reserves will always continue to be found. Ignoring problems – and the problems that will result are likely to be great (cp. Heinberg 2003, for example) – will not make them go away but only exacerbate their effects; yet the failure to even consider, let alone prepare for, probable and possible changes is currently a continuous pattern.

The more immediate, ecological problems of resources, rather than their finiteness, are the destruction or conversion (and attendant biological impoverishment) of ecosystems during extraction, and the environmental impact of their release or addition, e.g. in relation to the rates and kinds of processes in natural biogeochemical cycles.

Even human-dominated ecosystems such as agricultural ones are far from being closed, although in social-economic concerns related to them, they tend to be treated as if they were. Not only does (or did) their conversion from "more natural" systems incur an effect on ecology; but the introduction of additional and novel substances, too, is of wider effect: Nitrogen, for example, is quite naturally necessary for plant growth, and the technology for synthesizing ammonia for use as fertilizer in agriculture has been instrumental in modern agriculture. In the meantime, however, even the global nitrogen cycle is severely changed by human influence, the rates of nitrogen fixation and cycling by human activities being similar in size to those by biological processes (Mackenzie 2003: 192). The excessive introduction of nitrogen into agroecosystems, of which about half is estimated to seep away rather than actually serve as plant nutrient (Galloway 2003) is costly to farmers, affects ecosystems negatively (e.g. lower riverine and coastal areas, most prominently in the infamous dead zones in the Gulf of Mexico; but also forest areas, cf. Nosengo 2003). It has direct effects on human health as well, e.g. in negatively influencing rates of respiratory ailments, heart disease, and various forms of cancer (Townsend et al. 2003). Novel substances, even if relatively well-tested, too, can prove to ultimately produce undesirable results. The best known case are the chlorofluorocarbons (CFCs) that thin the ozone layer; more recently, concern has been raised over softeners used in plastics production which turn out to be having an hormonal effect (cp. Schettler et al. 2000, Myers et al. 2004).

The ecosphere limits that the issue of ecological impacts points forward to is a different matter from the procurement of fossil resources, with systemic rather than mechanistic dynamics. Simply to assume that human action were unnatural and therefore had to be removed (at least from wilderness), nor that it were always natural no matter what, nor that it were so all-pervasive that we would in effect be in control of the entire Earth, would help. Yet, the integratedness of culture/society in the ecosphere as an (ultimately) inescapable fact of life remains pervasive, and in order to achieve sustainability that truly is, requires that conservation and protection of the diverse living world – as the building block of the ecosphere that we depend on – be of the highest priority. The way these will have to be addressed for sustainability, however, is different from their usual connotation: Not the old model of environmental protection as cosmetic and end-of-pipe efforts, nor the conservation only of a few hot spots of biodiversity will do it, but the integration of human activities into the ecosphere oriented on its functioning, adapting to and coevolving with the biosphere's ecology, must be the aim.

Systemics

The need to address sustainability in terms of a "postnormal science" of systems (Kay 1999) becomes apparent with a consideration of the ecosphere as the system it is, which implies that its behavior, too, will be that of systems rather than of mechanic objects. In consequence, ecosystems will not be amenable to de- and reconstruction, but only to ongoing dynamic change of "natural" kinds, and their "state" could seemingly unexpectedly, if disturbance crosses critical thresholds in severity or cumulative effect, catastrophically shift to another degraded state (Scheffer et al. 2001). So, understanding the ecosphere as system, which is also a foundation of the "biospheric perspective" described above, implies that it cannot be treated as a static stage for human activities, firstly because its functioning requires the existence and continuing performance of its constituent parts and processes, and secondly because surprising change is always possible, and dynamic processes even the only way "natural balance" is produced.

Salient among those parts and processes is the existence and interaction in biotic communities of the biological diversity of species – which, it should not be forgotten, is a value in and of itself (at the very least, as the current end-point of millions of years of evolution), as well as a major value to humanity, in material, intellectual, and emotional terms. Even more importantly, the productivity, stability, and resilience of ecosystems, and in consequence their services, all are related to biodiversity. Even if their relationship is not always a positively enhancing one, integration taking heed of our biological inheritance and its (at least currently) best functioning in the structures we see in existence is a better advice than exaggerated belief in the potential of technological replacements or human "perfection," which is a particularly dangerous idea considering the lessons of evolution (see below). (However, the "integrated" addition of technology where it enhances human life is certainly better than calls for total abandonment, as well.)

At the same time, the systemic nature of ecosystems implies that some change – disturbances being the dominant shorter-term influences of this kind, and evolution coming to the fore in deep time – is indeed natural, and oftentimes even necessary for their functioning. There is, however, a guide to sustainability of the particular changes. For, as Daniel Botkin (1990) points out, the rates and impacts of changes that are truly natural, in the sense of their being in tune with ecological sustainability, and those that should be avoided because they pass thresholds of stability or resilience, can and need be explored and respected. Therefore, action in terms of integration of ecology and culture means "manag[ing] in terms of uncertainty, ... change, risk ... and complexity" (Botkin 1990: 155). With this approach, even ecological conservation (of a new kind, including human activity) is not only asked for, but could indeed be included in normal (integrated) economic activities, if – and only if – they too would follow the natural cycles, i.e. these "certain rates of change [that] are natural, desirable, and acceptable" (Botkin 1990: 157). Still, this yet again strengthens the requirement that an understanding of the empirical reality of the embeddedness of human life in the ecospheric system, and its existence only as constituted by living diversity, constituted the foundation on which theory and practice are oriented.

The integratedness of ecology and culture/society, ranging from the fulfillment of basic needs to psychological well-being and even spiritual realization, is another example for the actual systemic interrelation between most any issue, and simultaneously draws attention to the need to address the challenges of integration, of sustainability, in an appropriate systemic way. – Solutions will only be found by taking systemic connections into account, and utilizing the positive synergies (ensuring ecological functioning as well as improving quality of life) that this approach opens up.

The multiple causality of problems, their attendant complicatedness and mutual negative enhancement are rather well recognized, and coming to be so in issues that had been considered removed from the domain of nature in which environmental protection and conservation take place, e.g. in the relations between the environment and human health and sanity. The very history of human-nature relations is commonly described as the progressive emancipation of humanity from nature, but with the rise of losses caused by natural hazards and addressed in the context of global change, it seems that we may be entering a new era of stronger societal vulnerability to the vagaries of environmental conditions, and – for better if we followed routes towards sustainability, for worse if we do not – renewed stronger interaction between sociocultural aspects and nature, e.g. in health, safety, and well-being.

Yet, suggestions for a sociocultural transformation to sustainability all too commonly follow the fragmentation of disciplines and issues rather than the reality of their interconnection.

Linkages that are all but ignored in the wide field of the need to feed the world's growing population, for example, begin with the very dependency of population growth on food availability – the rise in human numbers that was made possible and indeed did occur together with rises in food production. This direction of the relationship could build (or be building) a vicious cycle of population growth requiring more production, more food allowing for further population growth, of utmost consequences, but was described only in a 2001 article by Hopfenberg and Pimentel. Moreover, population numbers and rates of growth are affected by a large number of additional factors: medical service (and its successes or the perception of them, e.g. in reducing infant mortality); knowledge, means, and support for family planning, or, to the contrary, support for large families; and even education and status of women, as well as quality of life. Considering these issues in a long-term and systemic perspective, and keeping in mind the existence of resource and systemic limits (to increases in agricultural productivity imposed by plant physiology and availability of arable land, affected by loss of soil quantity and fertility through current practices, and dependence on petrochemical and fossil energy input, as well as side-effects on other vital ecosystem services, not to mention that Malthusian limits must exist at some point), solutions to "feeding the world" then need to also address the cultural factors just described rather than focus just on increasing food supply. At the same time, we are finding that the decline in rates of world population growth is hailed as a success while countries in Europe where the age distribution is shifting towards more elder people than younger ones, and where population numbers are even starting to decline (which will ultimately be the case globally if the stabilization or decline of world population does occur as predicted recently) fail to consider changes in social organization, pension plans, and the like (which these developments will also make necessary), but suggest that the only solution were to raise birth rates.

In contrast to the negative feedback between environmental problems and the like, potential synergies in sustainability tend to be unrecognized and/or undervalued.

Again using nutrition as example, consider that traditional food systems and diets built on them are being given up around the world, mainly as they are seen as not being modern, not seen as successful and promising, and argued not to fit into international markets well enough (which are far from "free" because of the extreme distortions produced by subsidies, whether in the form of fossil fuel and petrochemical availability, or actual monetary support). However, they have been highly successful in satisfying nutritional demands in terms both of amount and composition – as well as providing expressions of local culture and identity in the form of regional/national cuisines; rather secure because of their diverse base of crops (including cultivated kinds and wild "emergency foods"; nowadays food security could be augmented by import); generally sustainable (at least in tendency and as long as population pressure remained comparatively low); and they would be "ecologically modernizable" (see Chapter 11).

One suggestion that remains to be made would be to also follow – "mimic" – natural systems in their imperfection, in other terms: flexibility, adaptability, and redundancy, which are not highly regarded, but well recommended from a connection of the systemic view (with its implication of changeability of conditions) with the evolutionary processes by which the ecosphere is adapting to and with changes.

Evolutionary Conservation

In addition to general and systemic ecological perspectives, the evolutionary-ecological point of view offers a valuable but under-valued input for the fundamental consideration of sustainability. In particular, its expansion of the range of reflection into deep time (as already used in the biospheric perspective), and of the mechanisms at work in it, offers a lesson on "success" and progress, and principles for co-evolution. The all-too-common failure of (mis)taking evolutionary theory for (a validation of) evolutionism, however, needs to be avoided.

As with the misconception in (Western) folk ecology that nature were a static, harmonious, mechanical stage (whether derived from religious sources seeing it as immutable and/or eternal creation, or from Newtonian-scientific sources describing it as a simple clockwork) rather than looking to the current state of scientific knowledge, so evolutionist views contrary to the implications of biological evolution abound. The comforting thought that humanity were the pinnacle that was meant to be – the splash of paint at the apex for which the Eiffel tower was built in Mark Twain's satiric analogue – was taken from religious sources, and toppled by Darwin's theory. Evolutionary/evolutionist "theories" predating and following its publication still exert strong influence, foremost among them the view that our (that is, usually, the Western capitalist, but Marxist thought follows the same schema) social organization and way of life were that to which all others "naturally" had to aspire, and which were legitimized by the Social Darwinist "survival of the strongest," if not a result of "natural laws."

In spite of all the controversy surrounding the biological theory of evolution, however, there are some insights that are very certain:

For one, that evolution is not a deterministic process aiming for an end-point, nor even knowing such a thing (but not unending in all cases, i.e. without extinction, either). Even within a species, the "fittest" that survive are certainly not characterized by their perfection, something else that evolution does not and cannot know (cf. Campbell & Reece 2001, Futuyma 1998) – for which of the multidimensional conditions of the abiotic and biotic, cyclically and every once in a while unforeseeably changing environment should an organism become "perfectly" adapted to?

The measure of success in biological evolution is quite simply (elegantly in a strange, "nature red in tooth and claw"-way) the reproductive success of a species and/or the individuals of a species exhibiting certain traits, the fitness in the Darwinian "survival of the fittest." (The importance of the survival of individuals led

Dawkins, 1976, to his infamous notion that the –"selfish" – gene's "point of view" were, as the genes were producing the traits that evolution worked on, an even better perspective for describing the evolutionary process than the entire organism.) Even so, all species can only exist within a range of population sizes dependent on their environment's carrying capacity, regulated either by mechanisms that make their population fluctuate around that (itself fluctuating) "ideal" size or by high rates of both reproduction and mortality (the k- or r-strategies learned in Ecology 101) – making human "success" so far appear both extraordinary, and highly risky. We can comfort ourselves by arguing that human beings, as elements of nature, could only perform "natural" activities, but spells of large scale population die-off and even extinction would be natural (but hardly wanted), too.

Secondly, however, evolutionary theory can be applied to human beings only with difficulty, in some cases such as the occurrence of lactose (in-/)tolerance in different populations it works very well (cf. Durham 1991 for an excellent treatise on cases of gene-culture coevolution, as well as their common independence of each other).

A lesson we could draw from our difference from all other species should yet be taken heed of: We have not passively become adapted by differential reproductive success for the most part (except with the prehistoric processes concerning lactose intolerance, skin color, and the like), but rather, we actively, creatively invent cultural means and ways of adaptation. In many cases, these constitute a decreased dependence on environmental factors, as far as there are ways around them – basic human needs such as oxygen, water, food, can or could be provided by technological means for a very limited time only, and even where they currently seem to be, their resources are more usually taken from nature rather than technically produced. In contrast, adaptation to temperature, for example, occurs mainly by way of technology (clothing, architecture, heating/cooling).

What this implies, in particular as regards sustainability, is that we (would) have the potential to come up with and prepare for evolutionary scenarios even before they have taken place. (In other respects, too, "naturalness" is not necessarily a good guide: Even a strong genetic determinist like Richard Dawkins has argued that we may be controlled by genes, but there were no reason why we couldn't rebel against that control. – That human parents can love an adopted child as their own, even if it does not spread their own genes as they should be "genetically programmed" to wish for, is an obvious case in point.)

The question concerning the transformation to sustainability will be whether we decide that it would be safer to plan ahead, heed the ecological connections already described and the linkages between culture, well-being, and ecology (that will be described in Parts 3 and 4), and start working on changing, or wait until change is imposed by environmental conditions in the form of "natural" mechanisms, thinking that we will be able to adapt to them, forfeiting the potential, or at least noble and sensible attempt, to let life, human and non-human alike, flourish.

From this point of view, the evolutionary-ecological perspective provides yet another reason for "fitting in," for proactive integration striving for as much environmental and species protection as achievable: Conservation, interpreted in evolutionary terms, "is not about stasis, but about maintaining the exciting, ever-evolving variety of life on Earth" (Knapp 2003). The changeability inherent in the workings of nature, and including systemic surprises, implies that we may yet see some charismatic megavertebrate species we would like to conserve go extinct (though the prospect will and should not keep us from trying anyway). On the other hand, the very certainty of change strongly requires that ecosystems with a diversity of species, many of which we may be less concerned with, be conserved so as to provide for a basis from which evolutionary processes may proceed, justifying biodiversity conservation from an evolutionary perspective (while the argument so far was on the ecological grounds that it makes up ecosystems, and hence their services).

A similar case can be made for society, as a diverse base of "cultures" (as ways of life, technology, etc.) not only "spreads impact" (Anderson 1996: 96), but provides for the cultural equivalent to what is, in hindsight, a preadaptation in biology. That is, technologies that are now considered quaint, ways of life that seem rather backwards, may prove to be necessary for further development/progress if or when conditions change. In evolutionary perspective, the attempt at optimizing ways of doing things which are currently dominant may – as they may not be representations of perfect, "fittest" culture developed by attention to their long-term potential, but rather of locked-in systems developed in specific historical paths (Mulder & van den Bergh 2001) – indeed be a way of producing problems more effectively. This is what we are seeing, for example, with the above-criticized "solution" of single-mindedly working to increase food availability without addressing the further population growth that could arise as a direct result of it.

A transformation to sustainability will, in conclusion, probably not only require efficiency gains, but include fundamental, systemic changes. Ideally, these should be instances of co-evolution between the (inextricably linked) social and ecological systems, preferably in ways that reduce the dependence of "natural" ecosystems and their services on human input in order that they can exist, develop, and provide us with services even when such input is unavailable (cp. Marten 2001). For the social/economic system itself, this means that a diversity of solutions oriented towards sustainability, fitting in with their respective surrounding environments and with global environmental conditions, both of natural environments and of people's cultures is most promising. They must not be based on current optima only, but need to follow an orientation on a long-term horizon, promoting "adaptive flexibility and evolutionary potential, which enables a continuous process of adaptive learning and a diversity of co-existing alternatives" (Rammel & van den Bergh 2003: 130). This hardly sounds revolutionary, but it runs counter to the current view that economic globalization (following a biosphere/global economy-people perspective not acknowledging physical-ecological reality) were the pinnacle of progress (even while it requires a further cultural-communicative globalization bringing home the point that humanity is "in this together"); but it is a challenge to the prominent quarrels over the one "silver-bullet solution" among scientists and environmentalists, as well.

The "social system," as this perspective also points out, is just like the ecosphere (and as part of it) made up of living beings, i.e. constituted by us and our actions. We will, by the very process of living and making a living, affect the future we and coming generations live in for better or worse, in any case; and we would be able to do our part in creating sustainable futures.

Chapter 7

Anthropological Contexts

Whereas the "natural" ecological-evolutionary dynamics are more difficult than commonly considered already, the even greater challenge of and for sustainability is "not managing the resources but managing the people" (Anderson 1996: 123). The biosphere, considering its own long history and development, could legitimately be argued to be sustainable – capable of continuing to exist and evolve into deep time – per definition. The ecosphere including humanity, and in a state that we like, depend on, and are by and large adapted to (having evolved in such environments), however, is quite another issue, requiring ways of life and thought amenable to sustainability, and the motivation to invent and follow them.

Discourses on ways to bring a transformation to sustainability about have so far focused on "the" solution, located either on the material side of politics and economics (and here debating whether legislation or free markets would do the trick), and of techno-science, or on the cognitive/mental side of world views, ethics, and philosophy. From an anthropological perspective, and including psychology, however, ways of action and of thought stand in a dialectical relationship of mutual interaction: Politics and economics are based on their own particular ways of seeing the world, currently as a particular, historically/evolutionarily "locked-in system" (Mulder & van den Bergh 2001) it appears, that may indeed require different frames of interpretation in order to even understand some of the concerns of sustainability, e.g. the non-utilitarian, emotional approaches. Science and technology, too, are based on their own cherished traditions which are not always supportive e.g. of the necessary, "post-normal" syntheses required. On the other hand, the nominal acceptance of environmental ethics – judging by the support for which, environmentalists have won the battle over the hearts of much the world's population already - obviously does not automatically produce more sustainable ways of life in practice. These relationships between thought and action have come to be implicated in the anthropological concept of culture, and its general holistic approach emphasizing the importance of both the material and the mental – culture as ways of life and as ways of knowing – (cf. Milton 1996) offers a way out of the quagmire of the dualist, dichotomizing rather than integrating discourses.

First of all, such wide anthropological approaches can be of value by providing a framework for consistently analyzing the relationships of the eco-cultural integratedness between humanity and nature, acknowledging their history and present, and potential synergies that sustainability could build on. Following a classification of human needs as frame of reference, this approach addresses actual necessities of human life (in their dependence on and relation with the environment) directly rather than by way of abstraction (i.e. only via economic mechanisms and growth). In addition, this scheme can be utilized to argue both from the global and universalistic perspective (of general human needs and ecological footprints), and from that of the particular modification of human needs produced by people's own agency and localization in particular environments, both ecological and cultural.

Secondly, the anthropological focus on culture, and on cultures in the plural, provides a necessary perspective on the diversity of world views, and particularly of culturally "fit(ting)" definitions of sustainability and a "good life." Besides informing the discourse with an understanding of universals and idiosyncrasies of human concerns and practices of how to live, and expectations and desires of what to want from life, this also points to cultural, social and institutional factors affecting resistance or support to different forms of living and knowing. Even more generally, it both questions and enhances the processes by which definitions of sustainability are produced and put to use, especially in the dual context of "sustainable development," where ecological foundations would need to lay the foundations, (Western) economic expertise is of largest effect, but cultural modifications that would make it a more people- rather than abstraction-oriented concept and practice are all but absent.

Thirdly, anthropological and psychological insights link these concerns of culture with (individual) action and agency, and inform promising approaches to building support for and shaping a (actually plural, diverse) "culture change" to sustainability.

The Framework of Human Needs

As anthropological research shows, the human condition is characterized by a distinct confluence of unity and diversity within and between human beings and cultures. The basis even of cognitive characteristics can be founded in our evolutionary history to a large extent, and therefore a universal feature inscribed into our genes (e.g. in the form of inherited predispositions and mechanisms as argued by exponents of evolutionary psychology, cp. Pinker 2002). Yet, at the same time, they can vary in accordance with the particular upbringing as members of a culture (e.g. as researched in psychological experiments, cp. Nisbett 2003).

The choice of food, about which it is always said that humans are just made to like particular kinds (sweet, fatty, calorie-rich) – and in America and Europe increasingly suffer health problems because of this innate like –, varies greatly between different populations in accordance with learned, culturally-contingent preferences (producing the well-known "ethnic" character in ethnic cuisines). Other basic – let alone not so basic – needs, too, are related to culture, as particular kinds of action and even thought acquired during socialization, modified by individual idiosyncrasies and decisions, but oftentimes unconscious, coming to the fore only in situations of "culture shock."

In keeping with the call for a stronger consideration of systemic connections, it must be noted that the various needs are, as a matter of course, related among themselves, as well. Nutrition and health, for example, are clearly contingent, e.g. as the amounts and types of food, their combination and preparation in certain ways (such as with herbs and spices that both taste well and have medicinal properties) influence physical well-being, and in their combination impart typical "ethnic" identity to dishes (which further relates food intake to social needs).

The common consideration of sustainable development based on the dominant economic view does not concern itself with such problems because human psychology, all the different needs and wants, are givens within its rational actor model, with everybody everywhere striving only for the maximum benefit in fulfillment of the same needs, by means which the economy provides. The relatively recent ("Reaganomics"-Thatcherian) idea of the trickle-down effect appears to combine with belief in the Environmental Kuznets Curve in this approach: growing the economy, everybody will eventually grow rich, and the environmental situation will improve (because there is both the money and the will to pay for remedial action). Here, all rational decision-making is based on monetary value, and money is seen as capable of providing for any and all human needs, life satisfaction, and control over resource use and even environmental/ecological factors.

The needs-based and person-centered approach is a conscious, active counter/contrast to this purely economic reasoning. After all, a major problem with sustainability is that ecological conservation nowadays typically takes the form of the "rich North" (which is industrialized and has seemingly lost most of the wilderness that would now be considered worthy of protection in the process) asking the "poor South" to forfeit development for the sake of humanity – while foregoing any changes to its lifestyle.

Basic human needs, however, are universally shared, and immediately dependent on the environment – and for the most part even the local environment accessible to experience in rational and sensual ways, except with elements of the globalized economy which tends to disengage people from place. Simultaneously, in the plurality of forms they, or rather their satisfaction (may) take, needs are a reflection of human and cultural diversity, particularly (but not only) once they become less basic and more contingent on cultural factors. Therefore, describing the integratedness of humanity and environment, ecology and culture, in terms of (the fulfillment of) needs means that the discussion of these issues is mindful of global differences, while it directly addresses human persons no matter where (with the necessary qualification that the long-term goal of sustainability can be easily overridden by more immediate concerns – but even this problem applies globally).

Moreover, this approach enables the development of an agenda that defines progress in more humane and sustainable terms: not simply as economic growth/development, but as the possibility for persons to work towards the achievement of a decent quality of life, characterized by the (potential) fulfillment of their needs, and preferably a "good life," as individually defined and/or supported by the wider culture. The qualification introduced by ecological necessities has to inform the definitions and does put a limit on the freedom of decision, as the environmental deterioration that has characterized progress defined solely as a measure of the economy, or a good life defined solely in terms of the possession of goods, ultimately works against the achievement of a good life for present and future generations alike, e.g. considering the interrelatedness between ecosystem health and human well-being. – What the individual definition of a "good life" could be retains a considerable breadth (probably more so than in the consumerist definition which mainly just knows material affluence), nonetheless.

Culture and Sustainability

The perceptions and interpretations which situate us in the world, one way of interpreting culture (cf. Milton 1996: 63), obviously modify the forms which the practical fulfillment of basic needs takes. In the context of nutrition, for example, they affect what is defined as edible in the first place, which are the preferred food choices, and in what combinations and preparations, etc. As a matter of course, these modifications not only influence other needs (as noted above), but they have effects on, and are in turn affected by, environmental conditions and actions in the environment.

Particularly as the needs concerned get less basic, even less immediately determined by biology, socio-cultural support for particular forms of learning and ways of life gain more leverage; they may even create perceived needs (a mechanism the consumer society is built on). Here, the effect of culture goes beyond the apparent diversity e.g. of "ethnic cuisines" to include a plurality of outlooks on the fundamental relationship between humanity and nature, as well as on what legitimately constitutes a good life (and a good environment, for that). Whether material affluence, emotional/experiential richness of life, or spiritual realization (or any or all of them) are considered legitimate and/or necessary foundations for "the good life," for example, largely depends on their being supported by the predominant cultural orientation.

The further importance in the context of sustainability is that "culture," in the sense of ways of seeing the world, shapes how the relationship between humanity and nature is perceived, and thereby provides salient orientations (as already used in the course of this work, in the section on "Human Perspectives"). Even this theoretical-philosophical concern is more complicated, certainly once actual behavior is starting to be considered as well, than commonly realized: The "ecocentric" view of a sacred, powerful nature sometimes upheld as a solution to the environmental crisis, for example, can be interpreted to mean that human activities could never destroy or despoil it, and be held even against all physical evidence (which can be seen in Indian attitudes to the river Ganges, which is considered holy and therefore, by definition, pure – unfortunately, the gap between cultural imaginary world and physical reality could barely be larger), effectively undermining the argument for protection. The same can happen if human activities are unequivocally interpreted as "natural," which is philosophically-empirically legitimate and maybe better than to consider them as existing in a sphere of their own removed from nature, but is nonsensical if they are not interpreted in ecological-evolutionary contexts. Culture, depending on its (evolutionary and) historical circumstances of development, can quite obviously include elements that are (and had to become) adaptive, but also elements that are maladaptive (but were not as effective as to be "weeded out" by exerting selective pressure), or neutral (or sexually selected, some biologists may want to argue). – For many (if not most) of them, it is quite certain that they were selected in the context of other cultural features rather than by "natural" factors and mechanisms (cf. Durham 1991).

The issue that comes to be of the highest concern in this discussion is that the culturally constructed meanings and valuations still play the paramount role in focusing attention on particular aspects (or selective denial) of the human-nature relationship, and on supporting particular ways of life; so that it is these culturally salient features (of life and/or of the environment) that will be of most concern to a given society. In result, as Rappaport (1984: 340) states, "it is neither objectively identified organic well-being nor ecosystemic integrity but culturally constructed value and meaning that ... human social systems tend, in the first instance, to preserve."

The task of a cultural theory for sustainability, then, has to be to link the analysis of culture with the ecological assessment of the positive or negative effect of the feature in question, finding out whether it is adaptive or maladaptive, contributing to sustainability or not (cp. Milton 1996). Of course, the challenge which binds analysis to activism would be to go from finding out what may serve as preexistent "cultural resource" for sustainability, to actually promoting deeper support for these positive aspects, especially in making them relevant and effective as orientations guiding life practice, and to change those cultural features which are of negative effect.

The explicit consideration of culture(s) in their behavioral/practical dimension, which is less emphasized in the term's recent usage, at this point becomes essential. – In spite of the discussion so far, "culture" is not quite as monolithic a system of ideas and rules-of-action as it is oftentimes taken to be. A better description would be that of a less-integrated system of relationships, consisting of a plurality of knowledges which become relevant only in particular contexts (e.g. of discourse or action). It may even be suggested that cognitive-cultural explanations (rationalizations) may oftentimes be invoked only after doing something simply the usual, conventional way, if pressed – by the anthropologist looking for the underlying cultural rules, for example – to elaborate.

The perspective of (historical) environmental anthropology suggests that societies that (by and large) did in fact achieve (a) sustainability in their relationship with the local environment managed to do so by combining the theoretical/cognitive, emotionally powerful, with the practical, encoding practices e.g. by religious explanation and transmission in socially binding ritual, as well as having had the relevant outlooks and practices become conventional and culturally supported (cf. Anderson 1996). Apparently, the achievement of sustainable ways of life is possible, but neither easy nor "natural."

The challenge in practice, therefore, is to establish linkages between thought, emotion, and action in the direction supportive of sustainability. This can entail: making the "cultural resources" of existing conceptualizations and practices which (in tendency) constitute sustainable ways of life more salient and strongly supported; establishing (support for) new such ways by showing them to be promising, positive approaches; or maybe even promoting alternative technologies and economics which fit in with already salient, less sustainability-promoting cultural orientations but are representing a step towards them in practice.

Ultimately, "cultures of sustainability" need to work in both ways, as cognitive orientation on and with which people are guiding ways of life towards that elusive aim (which will quite likely remain a minority issue requiring the input of ecological science used to develop the following), and as conventional ways of life and making a living (so that it ultimately does not remain a minority concern in actual life practice).

These suggestions are meant to be typical for the "positive ecology" approach as an attempt at moving beyond the merely oppositional approach arguing with difficult choices between conservation or development (which in effect means to resist change until the one perfect solution were found rather than moving in the right direction acknowledging that mistakes will be made but must be taken as lessons and should be following a "safe-fail" approach, i.e. be at scales at which their effects are not overwhelming but within the affected system's coping capacity).

Such an approach will be sensible, and even necessary, for psychological as well as anthropological reasons: The one because of the potential of such an approach to be motivational, supportive of human agency and inventiveness (compare below); the other because change typically is a minority cause at first, which can become successful only by initiating alternative practices that come to be "profitable" (not just in economic terms) and thereby more widely adopted, finally translating into wider (practical and cognitive) "culture change" as well. – The one without the other, practice without orientation, or views not followed up with fitting action, cannot go far; so both also become most effective, and in fact viable, only if they are both economically, in terms of the fulfillment of necessities for human life, and emotionally satisfying.

Defining what is sustainable has to be based on ecological analysis in its foundations (the understanding of which, without fear or favor, may easily be among the greatest challenges), but needs to bring it into fruitful discussion with "culture," so as to be(come) adapted to both kinds of environmental factors, natural and cultural. The pursuit of happiness, nowadays almost exclusively defined in Western capitalist terms as economic growth (in politics) and possession of goods (for the individual), is obviously a powerful motive; showing that and how a good life would be even more likely achievable (and in the end, only) with sustainability, thus requiring a re-definition of progress, could make this cause profit from that power. Utilizing such culturally supported factors rather than arguing only that they were or were not sensible in this context is necessary in providing both continuity of tradition and cultural identity, and a motivational positive argumentation. In other cases, too, this relates to the other/additional orientations that cultures provide, which are strong in the argument for autonomous definitions of sustainability and development/progress only rarely. Some indigenous peoples, for example, have come to strongly voice their own opinions on what they see as the right way for them to "develop," as do some religious groups (in Thailand and Bhutan, arguing from Buddhist backgrounds for example), as well as local groups and individuals, many of them from (or now classified as being in) the "anti-globalization" movement.

The hidden asset of defining the ecologically necessary demands and human concerns on which to orient sustainability only in general terms, leaving the working-out of details to the local arena – as is sensible since local natural and cultural conditions would need to be addressed – is that this approach does not disenfranchise people of their agency, but does provide a common theme, which unites the diversity of locally defined approaches at the abstract (but also relevant) global level of humanity-in-environment. Admittedly, it is only too easy to sell dreams of getting rich quickly, and more popular to engage in scapegoating rather than in accepting one's own role and responsibility, no matter how (seemingly) small. On the other hand, at least some people understanding the situation and eager to do something can be found just about everywhere. Sustainability analysis and activism should try to give them the theoretical and practical tools for setting out to work, rather than wait for a total worldwide consensus which the sheer diversity of human outlooks prevents anyway; and better still show that it is not (only) an attempt at "saving the planet," but an approach to living well.

The one anthropological consideration still to be addressed are just those perspectives on the role of and psychological dynamics in the individual, which usually get hidden behind the language of more abstract entities such as culture and institutions. However, as Anderson (1996: 123) reminds us, "societies and cultures exist only as emergent phenomena, without the flesh-and-blood reality of individual people ... they have their independent structures and dynamics. However, they cannot act or think – only individuals can do that."

Activism, Agency, and Change

Discourses on sustainability tend to focus on the large scale issues where both problems and solutions – or at least conventional approaches to them – are commonly seen to lie. Undoubtedly, such institutions as states or corporations can wield tremendous influence, which would need to be reoriented towards sustainability, defined in terms of ecology and concrete human concerns. Yet, because they are made up of, and made for, people (to represent them and their interests in state politics, or to distribute scarce resources by market dynamics), they come into the life of their own only from the basis of individual persons' interaction. And it will most likely be from that foundation that change towards sustainability will arise, if at all. Comprehensive approaches addressing both individual life, political, technological and economic workings, as well as social institutions or culture are needed. The aggregate of human behavior is what constitutes the foundation of these institutions, as well as of the diffuse environmental problems. So, large-scale change may indeed be required, but since the problems and their solutions are contingent on individual ideas and decisions, "the changes that are required to solve our environmental crisis involve changes in individual behavior" (Zelezny & Schultz 2000: 366).

More importantly for the work on a transformation to sustainability, it is in the innovativeness and entrepreneurship of individuals and groups committed to do their part that change originates, and it may well be the effect of such change that further support is raised. As James Gustave Speth (2004: xii) notes, "The current system of international efforts to help the environment simply isn't working. The design makes sure it won't work, and the statistics keep getting worse. We need a new design, and to make that happen, civil society must take the helm." Choices, example, and motivations of individual people, therefore, are of paramount importance.

Environmentalism as it has mainly been arguing to date, with its focus on the bad news and dangers and appeals to sacrificial altruism contributes to the feelings of well-informed futility and learned helplessness (cp. "Sustainability Paradox"). In effect, this approach contributes to the very de-motivation and denial that hinder understanding of the argument for, and practical effective steps toward, sustainability (cp. Kaplan 2000).

Activism, for example, has been instrumental in building awareness of issues and support for environmental and development organizations and their causes. Its routine adversarial approach needs emblematic problems and problem-causing "villains." Where such can be found, it can be highly effective for some time and for bringing about limited, particular activities, e.g. boycotts and changes in business practice, providing one way to influence the activities of corporations or states.

With other – most – of the challenges of sustainability, however, both perpetrators and culprits are simply us human beings, whether the problems are caused out of necessity arising from poverty, or out of unthinking consumerism. Here, adversarial approaches focused on guilt, arguing for changes that (at least in the way the argument goes) might help the world but hurt one's own lifestyle, run up against strong psychological tendencies, e.g. of denial of outcome severity, stakeholder inclusion, and self-involvement (Opotow & Weiss 2000). This is particularly salient as the problems are caused by a modern (Enlightenment, but not "ecologically enlightened") project that is not aiming for ills, but rather for better human welfare, or at least, in the capitalist (American) dream, for the potential to achieve happiness by dedication and hard work (or more recently, by becoming famous and thereby rich). Environmentalism, however, by arguing that things will have to change (be given up even), without putting them into the context of alternative, environmentally and humanely better solutions appears to be working against this very hope, fostering the perception that it is better to live well now because things will only get worse, and not much could be done about it anyway (cp. Kaplan 2000).

Two major issues that have been under-acknowledged, but could be used in providing motivation are the inherent human need for perceived self-efficacy, and this same hope for a better future: The already often-enough mentioned foundation of positive ecology needed to lie in its aim of providing empirically justified and motivational, "understandable and attractive" (Csikszentmihaly & Seligman 2000), approaches through its combination of the argument for the (long-term) necessity of ecological sustainability with the interrelation of human well-being with the environment, and in presenting solutions toward sustainability that exhibit positive synergies ("multiply desirable choices," Kaplan 2000) between the conservation of ecological processes, the fulfillment of human needs, improvement of human well-being, and potential for humanity.

The pursuit of happiness, in a psychological perspective, not only points to relations with the environment in issues of biophilia and well-being, or human (higher-level) needs and experience in and of nature (which will be considered in more detail in a later part). It is also linked to personal agency in various ways, e.g. as a counter to consumerism, for improving the sense of personal efficacy, and recovering individual agency. For example, more immediate personal satisfaction can be drawn from possibilities for working in parallel with one's convictions and wish (for oneself and one's children) to live in a world worth living in. The understanding that some needs can not be satisfactorily addressed even by excessive material consumption, which is also unsustainable, could play its role in furthering the positive notion of action for sustainability, but only if it is shown to be "multiply desirable" (for ecological reasons, human and individual well-being), not if it appears to be just another case of environmentalists telling us that we have to do without this and that.

People also generally want a sense of having a grip on their lives and the world – and preferably towards the better, – which personal involvement in decisions and actual work, together with (informal) education on current issues and scenarios for the future by (or rather with) experts could provide.

Not all people would initiate change, whether in their own lives or aimed at the wider world, however, nor even exhibit ecological understanding or "profits" from the experience of nature. Therefore, providing a guide and explanatory tool for those (few) who would has to be one of the aims of a "positive ecology." This approach also offers itself for the "marketing perspective" suggested by Gail Whiteman (1999) that would aid in promoting the transformation more widely. Theoretical guidelines, e.g. rules of conduct to orient on, and practical examples are necessary, as well – with every small step recognized as such, the next ones are easier, and more people may decide to follow.

Over time, the initiation of change by individuals and small groups can thus start to gain wider influence, at least once the alternatives come to be seen as promising, and to be put in practice. Whether agency is recognized as a motive force, and individual/community empowerment as a helpful development or not, is just like "traditional" cultural support for or against sustainability related to cultural orientations: the predominant "culture of fear" (Glassner 2000) in the USA (and similar European views of a dog-eat-dog world, "homo homini lupus," for the classically educated), for example, make people be seen mainly in terms of danger and a need for distrust, which hides well-working community initiatives which are already under way and draw a brighter, more hopeful picture.

It is highly important to note that this process of introducing change is explicitly meant to include (together with the personal and local-community initiatives one would more immediately think of) initiative in government and especially in the private sector. The development of practices which profit from working with rather than against nature by innovative entrepreneurs would exert tremendous influence. In addition to the generally wide influence of business, which is in many cases stronger than that of governments nowadays, the likely susceptibility of the private sector to both social and economic-competitive pressure, and openness to arguments for profitability with sustainability makes it a necessary and promising, if reluctant, partner (Daily & Walker 2000). The verdict on whether or not more profit can be made with sustainability in the short-term is, admittedly, still out (but see Lovins & Lovins 2001 for an argument towards that view). Business-as-usual may appear to be better on this count as yet, but only in the absence of stronger pressure for an orientation other than on growth and shareholder value alone, with lack of feedback on the social and ecological impact of its practices (rather the reverse because of false systems of measures and rewards), and missing realization of its dependence on the ecosphere. In the long term, as some sectors of business are already coming to realize or even experience, however, sustainability is quite certainly the only way, as "shortages of ecosystem services are likely to become the limiting factor to prosperity in [this] century" (Lovins, Lovins & Hawken 1999). This makes it all the more obvious that "green entrepreneurship is not only important because it provides new opportunities for the nimble firstmovers who identify and exploit such opportunities but also because it has the potential to be a major force in the overall transition to a more sustainable business paradigm" (Schaper 2002: 27).

Sustainability-oriented businesses that prosper by doing so, individual and community examples of widely – not just, but also economically – "profiting with nature," are the ideal illustrations of the necessary, positively-oriented approach, and the ideal examples for raising awareness and motivation for it. – Ecosphere structure and function will only be conserved, and human needs and well-being addressed in conjunction with them, if their interrelations are understood, and if ways to live sustainably are developed, seen to work and fit with personal aspirations and social/cultural demands and support, and eventually coming to be the supported and/or demanded ways of life themselves.

What finally needs to be realized is that "the solution" to sustainability can only lie in a comprehensive approach. As this discussion aiming for a synthetic theory incidentally shows, it may not have to imply the development of a perfectly fitting synthesis between issues, but it does mean that the disciplines and issues that need to be considered, do all need to be considered. Ecological-scientific input will be essential, and should finally make itself understood to be so, e.g. by education, participatory scenario-planning, or even the suggestion of what economies built on ecological foundations would look like. – Ecological principles are less understood, but no less real than the laws of physics; they are easier to break and consider unreal, too, but only for some time, which makes it all the more necessary to show that they do not mean that we have to give up hope for the future, but that setting out to work for a transformation to sustainability, with and in nature, far from hurting human welfare, can actually, and ultimately only, ensure human survival and well-being. Just as the laws of physics do not prevent a staggering diversity of life, however, so the "laws of ecology" definitely do not prevent a diversity of cultures, both as cognitive constructs and as ways of life, and a diversity of individual outlooks. This diversity, as human needs, ways of (making a) living and aims in life in general, needs to be taken into consideration, e.g. by stakeholder involvement or through community empowerment, promotion of initiative, at best in dialogue with science. This way, it may be hoped that the environmentalist/real world-perspective would eventually gain wider currency, but would not need to be accepted by any and all because successful ways of life and of making a living are based on its principles anyway.

Chapter 8

"Integration"

To adequately address sustainability as suggested here, two basic steps are in order:

First, to consider the intrinsic relationship between humanity as part of the ecosphere, that is, the fact of our deep dependence as well as "integratedness" within it. The opposite, an intrinsic relation between the ecosphere with human beings does not necessarily hold true – the human species did not exist for all of the biospheric history, indeed for least of it, after all – but something would be missing without people. And a perspective not addressing human concerns would hardly be the motivational one here sought after.

Secondly, then, the two sides need to be considered as one, with a view towards the potential for the continuing existence and development of both people and the diversity of non-human life. This requires that, since we are the more active, and especially the conscious part, we worked towards this goal.

The basic notion of (the) "integration" necessary for sustainability is formed by the suggestion of focusing on actively making human activities fit in and work with ecological patterns and processes. This, in order to move towards ways of fulfilling human needs and of enhancing the potential for human development and progress so as to open up the future by also promoting ecological sustainability rather than decrease future potential by denying the need for taking global ecology into consideration. Such an idea is the guiding orientation and yardstick variously proposed in "Natural Capitalism" (Hawken, Lovins & Lovins 1999), industrial ecology or biomimicry (Benyus 1998), the emerging field of ecological design (Orr 2002), and so on. "Natural Capitalism" is particularly noteworthy for the argument that humanity were moving into a situation in which ecological patterns and processes (resources, especially of the living kind, and ecosystem services, mainly) set the limits to development, not the means of utilizing/harvesting them, so that a change beyond the industrial revolution to a "natural capitalist" one is coming to be required.

In contrast to currently popular notions of stewardship, and more especially ecosystem or biosphere management, however, it is more inclusive, comprehensive – not least because these tend to forget the need to manage societal activity. Moreover, notions of biosphere management, let alone geo-engineering exhibit a fair amount of hubris considering the likelihood of unforeseen effects, particularly when, e.g. sustainable rates of harvest may be based on – imperfect – assumptions of natural stability, for example (cf. Botkin 1990). Rather, it follows the notion of Gerald G. Marten (2001: 169) that sustainable human-ecosystem interaction is not one that is (only) managing ecosystems in what should be such a way, but rather one focused on enabling the continuing functioning and provision of ecosystem services (by resilience) without (an inordinate) dependence on human involvement. Also, it follows the need to manage human – our – demands on the ecosphere. Vaclav Smil is one of the few authors to have clearly called for this approach, arguing that the goals of lowering consumption while enabling higher quality of life for the poor "demand that we manage our societies as never before, that we develop a new economics consonant with the long-term maintenance of biospheric integrity" (1994: 109).

"Integration" must not, however, be mistaken for a simplistic "back to nature," as it applies to human activities ranging from subsistence (eco-)agriculture (cp. McNeely & Scherr 2001) to industrial processes (cp. Benyus 1998, McDonough & Braungart 2002). – And while it were dangerous to assume that humanity were able to replace nature with technology (let alone that it should), at the very least since such a situation would not provide any safety net, integration would – and would indeed have to – coexist with and enhance human inventiveness (including in technology) but with a background founded on ecological considerations (and probably, admittedly, asking more deliberately whether every innovation truly provides a value other than its novelty).

Integration, as an active process of making human activities fit into and mirror ecological cycles, would contain the idea of environmental protection – from widespread alteration of natural ecosystems, introduction of novel substances, etc. – (thus including the precautionary principle) in its very workings. It should also be going hand in hand with biological conservation. In some cases, such as with biodiversity hotspots, conservation that means keeping human beings out of core areas may be preferable (from the biologist's perspective, at least), but generally human beings' existence in such areas should not be considered unnatural or in any other way illicit, yet (re)oriented in such ways as to enable ecosystemic functioning and the flourishing of biodiversity together with that of human inhabitants. This orientation applies to the industrially developed "North" where conservation has commonly been seen as an issue removed from people's (certainly "Northerner's") lives and to be performed only by such removal (with conservation only taking place in, and in the form of, national parks) just the same as for the "Third World" where it would be needed to enable a development with results that are not undermined by its environmental and social impacts. Sustainable conservation and restoration of nature even around and with human inhabitants hence is to be an element of integration, as suggested by Rosenzweig (2003) under the title of "reconciliation ecology." Finally, it needs to focus on how to make economic activities oriented on and mimicking ecological processes, e.g. by a stronger focus on services and cyclical flow (of all substances used) and by design for efficacy (cp. Hawken, Lovins & Lovins 1999 and Benyus 1998, McDonough & Braungart 2002, respectively), as well as to consider, more generally, what the appropriate means for the fulfillment of human needs are (which may not always be solely products).

Most importantly, translating the argument for integration into practice takes a realization of the need to work with rather than against nature, and with the best of our current ecological knowledge – first of all, therefore, of the need for living and making a living in ways that are in conjunction with the qualities inherent in complex systems such as the ecosphere and (including) human culture. The general principles this implies are the ecological-evolutionary systemics and patterns, as well as considerations from anthropological points of view derived from the discussion in Chapters 6 and 7.

The most obvious changes these would entail are:

1. a new orientation on natural capital and new forms of (ac)counting progress in ways including what are currently externalities such as conservation and restoration of natural capital, ecosystem services and processes, as well as immediate satisfaction of human needs, and if possible, indicators of well-being and possibility, (instead of or in addition to subsuming them under the proxy of economic growth);
2. acknowledgment, and the utilization as guiding principle, of the necessity for conservation/restoration (sustainability) of natural capital in considering (empirically, for current and developing situations) long-term viability, with
3. direct, local/community satisfaction of basic needs going hand in hand with development of appropriate special goods for wider-ranging trade, as a way of combining ecological orientation and advantages (e.g. security through lower dependence on transport) with advantages – and simply reality – of trading systems, as well as deriving further (e.g. food) security from potential trade in such basic goods; this in combination, but not in the form of a only virtual viability achieved only by expansion that cannot continue indefinitely.

Integration cannot be solely based on global, general scientific, principles being translated into local action; the second pillar on which it needs to be based, considering the importance of sociocultural dynamics within it, will lie with a "fittedness" in terms of culture(s), as well: At the least, a translation into vernacular languages and salient ideas to further understanding and, hopefully, support (cp. Thaman 2002). The role of cultures, and of anthropology, is of the greatest because it appears that ideas of or helpful for sustainability can be found everywhere, for one, and cultural diversity, e.g. of ideas of what makes a life well worth living provide one of the best counters to what Michel-Rolph Trouillot (2003: 139) describes as the "secular religion ... that constructs economic growth as the ultimate human value." As he goes on to argue, "we [anthropologists] owe it to ourselves and to our interlocutors to say loudly that we have seen alternative visions of humankind ... and that we know that this one may not be the most respectful of the planet we share, nor indeed the most accurate nor the most practical. We also owe it to ourselves to say that it is not the most beautiful nor the most optimistic."

"Spiritual ecology" as a "complex and diverse arena of spiritual, emotional, intellectual, and practical activities at the interface of religions and the environment"

(Sponsel 2001: 181; cp. Ch. 18), for example, is among the most interesting features – not because the silver-bullet solution lay with a re-enchantment of nature, nor even because it were relevant to any and all persons, but because it is one of the most outspoken of perspectives in pointing out that there is more to human life, and relevant in all cultures and many a life, than possession of material goods and the pursuit of status and money.

The description of actual integratedness, and of potential pathways toward an integration of human activity to work with and within ecological-cultural local environments and global ecology in regard to particular issues (utilizing the framework of human needs) will be addressed in the further course of this work.

PART 3
Fundamentals of Integration

Basic Needs and Ecology

Understanding – and making understandable – the perspective of sustainability as implied in positive ecology requires, first of all, that the inclusion of humanity in nature is considered. The various ways in which human life depends on and is related to the ecosphere, landscapes, species, etc., are seldom comprehensively described. – Natural science is concerned with nature but mainly excludes humanity (except for human beings as physical bodies) from its consideration, economics focuses exclusively on the material aspects, but usually not on any ecological foundations, the humanities, in contrast, are focused on the cognitive/ideational aspects (as far as it is concerned) quite outside of nature.

Human life, as is obvious to the non-specialist, and should be clear in objective views, includes reason and emotion, life practice and world views. So, in contrast to Wilson's (2002: 40) note that environmentalism "is not yet a general worldview ... compelling enough to distract many people away from the primal diversions of sport, politics, religion, and private wealth," sustainable cultures would actually need to include the whole range of concerns: Human needs, in their simultaneous universality and personal, cultural, or local idiosyncrasy of expression, and in their wide range from the utilitarian to the cognitive and possibly even spiritual, offer an excellent (even though little utilized) framework for describing the actual relationship between "primal diversions," not always diverting but primal concerns, cognition, and nature. In what ways the satisfaction of needs, and the advancement of the human prospect and well-being can be attained with sustainability – with an orientation on "integration" – is also well addressed in this context. Indeed, it needs to be considered simultaneously in order not to fall into the trap of the "ecological marketing mistake" as described in the introduction, of ostensibly providing the foundation without which sustainability were not possible, but leaving the development of practical solutions to those ignorant or in denial of those foundations (and then complaining about it).

The basic human needs can be classified following Anderson (1996, who is building on Maslow 1970), and connected with Kellert's (1993) typology of human valuations of nature to provide an excellent framework for this discussion. Nevertheless, their mutual interrelation, as well as the special (multi-purpose if you will) ways in which they are satisfied make any listing somewhat arbitrary, follow separations that are in fact showing considerable overlap:

Water, nutrition, temperature regulation, physical safety, sleep and arousal, health, as well as sex (with its accompanying result of reproduction) are apparent basic needs; utilitarian, dominionistic, and negativistic valuations (the use of nature as resource, the wish to control it, and aversity to features perceived as dangerous such as snakes or spiders, respectively; Kellert 1993) are closely associated with these needs. – Water, for example, is mainly wished to be used and controlled, but as a carrier of pollution or taking the form of natural hazards also feared as danger to health or safety, yet further related to other valuations, e.g. as a symbol of purity. Temperature regulation and physical safety (from vagaries of environmental conditions, as protection from the elements) are the best examples for the special human way of satisfying needs, as these problems are taken on by manufacturing clothing and constructing (built) environments, resulting in larger changes in the natural environment than may be necessary in strict terms of biological survival. But, of course, human life typically is about more than just getting by and reproducing. This approach of not being physically/biologically adapted to the majority of environments we occupy as much as constructing the conditions we need within them (satisfying other, e.g. social, cognitive needs and likes at the same time), is exactly what appears to have made the human species so successful. (On a biological note, it should be added that other animals engage in "niche construction" in which their activities move environmental conditions in a direction amenable to their "lifestyle"/ecological niche, as well, albeit to more limited extent.)

Interestingly, in another relation to the rest of the animal world, but extended in homo sapiens because of the rising importance of psychological and sociocultural factors, a need for perceived control or self-efficacy (the feeling that one is sufficiently in control of one's life) is highly important for health and well-being (cf. Anderson 1996: 87f., who refers to Langer 1983, Schulz 1976, Bandura 1982; the writings on the opposite, learned helplessness, are also highly instructive, e.g. those by Peterson, Maier & Seligman 1995), but commonly overlooked in sustainability (both environmentalist/ecological and development) discourses.

As social and in both action and thought creative animals, social needs and different aspects of "needed" cognition are basic features of human mental and material life, as well. Among these, there exists "human beings' inescapable and largely unconscious appreciation of the inherent structure of biological reality" (Berlin 1992: 3) which is intimately related to Wilson's (1984) concept of "biophilia" as the human "need" to affiliate with life and life-like processes. The different valuations of nature as classified by Kellert (1993) correlate with these various needs, as well:

The humanistic valuation (e.g. keeping pets) reflects one element of the social need as applied to nature. Naturalistic, ecologistic-scientific, moralistic, symbolic, and aesthetic valuations are associated with the cognitive needs (and impacting on perceived control and social needs) which are more and more contingent on culture- and person-specific factors, and rather less directly – although they still are – related to environmental features, as well as farther removed from the more common utilitarian-economic needs and interpretations of value. The spiritual dimensions these latter types of valuation point to are even defined, e.g. by Cooper & Palmer (1998: ix) exactly by their not being instrumental or pragmatic outlooks on nature.

Chapter 9

Conditions and Contexts

As noted in Part I, world views/cultural orientations provide one very basic and strong background against which the notion of sustainability is interpreted. Such "imaginary worlds" have been highly instrumental in guiding human behavior, but may be relatively far removed from empirical, ecological reality. Nevertheless, people tend to protect them against contradictory information as long as humanly possible – whether it were economists/managers who would not want to consider the environment as the ultimate base of the economy (and indeed life) that it is, or radical environmentalists who, having decided that the economy were always the culprit, would not want to see that trade has always been an element of human culture.

There are other contexts to sustainability, located in a middle range between the rather philosophical sphere of world views (which, in their scientific discussion are represented by a further abstraction from their lived-in form, which may have a rather less reflected, subconscious character belied by the reality and clear definition that the term seems to imply) and the most concrete level of human needs. In the current situation, the issues of population growth and/or numbers remain among the most important contexts. So are the recent urban condition impacting on sustainability (see Chapter 13), and the challenges of the energy requirements of modern society which only form an aside in this work, both because they are a condition among the fulfillment of different needs but not a need in themselves, and because Vaclav Smil (2003) has addressed this topic much better than could be done here.

The Pivotal Issue of Population

The human need for sexual activity – which is truly a need, or maybe better described as a biological urge as one could live without it, albeit apparently with difficulty – is best taken out of the following discussion (of what are strictly needs). Yet, it has to be considered as an influence on a major condition which sustainability will or will not be developed under, that of human population.

Nature, as environment, plays little direct role in the context of sexuality as a need (although it does not stand to forget that romantic, or even erotic, settings may have a lot to do with natural settings and materials). Nature as inner essence, as a matter of course exerts a strong hold over us where this urge is concerned, possibly even the strongest. At the very least, sex is the one "primal diversion" that always seems to be of interest, and wrapped into sociocultural views and sanctions to the highest degree. (Tellingly, even proponents of a transfer of human consciousness into computers in/as a posthuman, technological stage of our development, no matter how much they dislike the trappings of a biological, physical body, invariably seem to still include sexual activity in their visions of this future.)

Reproduction, the natural correlate of sex – even though the religious right should take note of the equally natural human interest in making this fun without that consequence possible – on the other hand, is not an individual human need. Yet, it is obviously necessary for the continuance of species and social groups, so that it does not come as a surprise (even from biological and anthropological perspectives) that the emotional bond to (one's) children is generally very deep and quick to develop once that consequence does indeed arise, and that individual or family decisions, social conventions and sanctions as well as ideas surrounding reproduction should be salient elements in any social group's culture. And, as has come to be well realized because of its (earlier) wide treatment in the (environmentalist) literature, the issue of population numbers and growth, and the accompanying impact on the environment and produced by increased consumption of resources are a concern in the context of sustainability indeed.

The contemporary "human domination of ecosystems," which implies that "humans are stewards of nature, whether we like it or not" (Sanderson et al. 2002), is a direct result of human numbers, together with technological capability and with levels of consumption. Unfortunately, concepts such as human ecosystem domination, but even stewardship or planetary management, have a tendency of being interpreted as meaning that we had total control over ecosystems and were thus standing above and apart from them. Nevertheless, ecological "laws" necessarily hold their sway over us as the physically existing beings we are, and various natural phenomena are adamant to our attempts at influencing them – the influence taken in the form of human domination most typically, and unfortunately, is a destructive one, at best raising productivity of the one resource we are interested in, but decreasing ecosystem services and biological diversity.

In fact, because of the relation between mere population numbers and environmental impact, these effects of human domination would not be seriously dampened even if humanity as a whole would (and could) decide to leave the wilderness currently remaining alone. Actually, this would not only not be enough because of the reduced size of such areas, and indeed the virtual non-existence of wilderness as environments without any human presence or interference. This approach, e.g. of classical conservation, even tends to be an outright dangerous way – taken by itself, that is, though it is one approach to sustainability together with restoration and reconciliation ecology (cf. Rosenzweig 2003) – because of its attendant assumption that these protected areas were all the nature worthy of protection, and taken care of by its designation as protected, leaving all other areas to be utterly changed as desired for short-term gain.

Biodiversity, ecosystems providing many services, in short: nature, but including a diversity of varieties and landscapes created with human influence have been existing and can continue to exist with us: some human dominated ecosystems, of course, are better from a sustainability or from a conservationist perspective than others (and these two, no matter how related, can differ considerably), providing for biodiversity and ecosystem services, but also for human needs. In some cases this even goes to the point where landscapes are not recognized as being cultural (or in other terms, historical examples of human-dominated ecosystems), but are most commonly considered to be natural; and even in other cases, e.g. considering cities, the ultimate in human-dominated and even human-created "ecosystems," some surprisingly large biodiversity can remain, finding new suitable niches.

It is not enough to simply assume that nature were adaptable anyway, however. A wholly technologized planet, even plus the species that manage to get by on it, may not make up an ecosphere that sustains us, and would definitely be a fiasco in any non-utilitarian terms (and possibly even considering human psychological health). A more natural-like ecosphere managed in its entirety, similarly, would probably be doomed for failure. In this case, the ongoing need for human influence, the variety of management aims, systemic surprises, etc., that would impact on it would likely lead to problems. So, less of an influence – as well as of the need for it – would be a preferable approach (cp. Marten 2001).

Yet, to get back to this point, even if human domination of many ecosystems will not go away and "integration" presents a major challenge, it appears possible to change both quantity and quality of human impact so as to co-exist with biodiversity and fit in with the structures and functions of the ecosphere, which sustain human life after all.

Conventionally, ecologists would argue that there is a certain population size that can be sustained within a certain region/ecosystem, judging by the resources needed and provided. With humanity, this carrying capacity is all but meaningless because technologies/techniques and (consumption) choices can change it strongly, and because "global economy people" draw on resources of the entire biosphere, not just the location in which they live. Related concepts, however, can be meaningfully employed to address the relationship between human population numbers, consumption, and environmental factors, e.g. the "human (ecological) footprint."

This concept (cf. Wackernagel and Rees 1996) provides a measure of consumption of energy and material resources (and including the sinks necessary for wastes, carbon emission, etc.) of a person or population, equivalent to Ehrlich and Holdren's IPAT formula, i.e. the relationship that human impact equals population times affluence (consumption) times technology (efficiency). However, the human footprint represents these factors in terms of biologically productive area (of land, e.g. cropland, forest, etc., and sea), giving a graphical measure, and one that points to the dependence on a variety of ecosystems. Comparing the "footprints" of people in different countries, moreover, shows the differences in relative impact (although it should be noted that this is more telling for overconsumption than non-sustainable underconsumption, i.e. underrepresents where natural capital is destroyed only just for getting by). When setting levels of consumption for the entire human population against the total area available on this planet, it can be shown what level of consumption would probably be globally sustainable (only "probably" because many difficulties preventing exactitude naturally occur). The additional possibility of charting levels of human influence onto world maps shows which regions are most changed, and which ones are "the last of the wild" (Sanderson et al. 2002), helping to guide approaches to conservation, e.g. by telling us which areas are best suited for classical conservation, and which require restoration or "reconciliation."

In spite of the inapplicability of a strictly determined carrying capacity for human population, matter of fact remains that some upper limit to size and resource consumption must ultimately exist. – Wackernagel and Rees's research, for example, suggests that the spread of current levels of consumption in industrialized countries to the world's entire population would require three to four entire planets Earth, so this kind of consumption cannot be sustainable; sustainable lifestyles in terms of ecology and social equity will therefore imply changes in both "North" and "South." Whatever the measure used and the number given, a higher population size obviously translates into fewer resources per person, and therefore more need for caution. Too high a number of people will either be reduced by voluntary, global activity, or "naturally" imposed by social strife and lack of ecosystem services and resources (maybe also disease). A lower population size, on the other hand, would quite certainly make the achievement of a higher standard of living, and of sustainability, easier. Or, put the other way round, even where population growth may not be the most important issue, "few if any of these problems [(of) expanded food production and better resource management] will be resolved through rapid population growth. They are the context on which this growth will be imposed" (Preston 1994: 9). Unfortunately, the standard operating procedure so far is that initiatives at arresting population growth are the very best to be had, and even this issue is rather avoided, whether for religious (dislike of family planning and measures), sociopolitical (failure to even want to find alternatives to the current generation contract based on unlimited, continuous population growth), or economic-political reasons (denial of unsustainability, and possibly even hope that more people means more customers and more, cheaper labor). Recent articles actually arguing for the preferable or necessary reduction of population size (as opposed to mere stabilization of population growth and/or size) are even fewer (cf., for example, Smail 2002).

One note of hope, connecting "positive ecology" in particular with the context of population lies in the apparent relationship between a good life and the number of children: Ostensibly, when basic needs are satisfied, and/or hope for the future relatively bright, knowledge of family planning methods available and women empowered to use them (cp. Pirages & DeGeest 2004: 51), and/or chances that one's child or children will survive and not be needed to provide for the aged parents, reproduction becomes less of an issue.

With that, however, the transformation from social-economic systems based on growth of both the economy and the population size to different systems comes to be of rising importance. Japan and Europe already are getting into this situation (particularly as regards population), but so far do not manage to consider any responses other than calls for higher fertility of their populations.

Chapter 10

Water

The need(s) for food and water would appear to be straightforward examples for the dependence of humanity on ecological conditions, as well as for the ecological changes wrought by human activities with the purpose of satisfying these needs. The intimate relationship between the settling of human groups and aquatic environments – such that virtually all of the "great civilizations" arose close to rivers or other convenient (e.g. regularly replenished) sources of water, and found ways of utilizing them effectively – testifies to this dependence. That water is quite simply "the stuff of life" without which no life as we know it could exist, however, makes it a difficult example for the positive aspects of sustainability, especially since it is, at the same time as its being an essential resource for survival, the most undervalued one, and not usually considered from a systemic, ecological or eco-cultural, perspective. Moreover, its utilization as a solvent and drain for pollution (whether metaphysical, bodily, or industrial), and its capacity for acting as an element of danger, e.g. as vector of disease and natural hazard, make water appear in a dual nature, both essential and ambiguous. Thus, the concerns over water span both the issue of quality, and of a – sufficient, not destructively high or low – quantity.

The human biological need for drinking water is quite straightforwardly set at an amount of 2.5–3.5 liters per day, with some margin depending on exertion and environmental conditions such as temperature and air humidity. Yet, even just for survival somewhere between 20 and 50 liters are necessary to provide for cooking water, washing, and hygiene, as well. The additional utilization of water for sanitary purposes, irrigated agriculture, and increasingly in industrial processes, lifts the amount of annual renewable water deemed necessary per person to 1,700 m³. Regions where water availability is below that level are defined as suffering from water stress. However, the rise in population and concurrent increase in water use have a strong bearing on the actual level of water stress in the various regions.

Water Everywhere?

Earth is rightly considered the planet of water, only here (amongst the planets in the solar system) does water naturally occur in all three states of matter, and even covers 71 percent of the planet's surface. Of all the water present (using the data from Shiklomanov 1993), however, by far most is salt water contained in the oceans (~97.5 percent) and therefore usable only with a high input of energy for desalination; the remainder is freshwater of which most (1.75 percent of total water, 68.7 percent of total freshwater) is in the form of permanent snow and ice, which is, as it is mainly located over the poles, again not usable. Only ~0.77 percent of total water (30.38 percent of total freshwater) is potentially usable, but even of these the majority (30.06 percent of total freshwater) is locked away in groundwater which is recharged at relatively to very low rates. This leaves only precipitation over land as the main source of renewable freshwater supply, which, to complicate matters more, is very unevenly distributed across Earth's regions, and varies in seasonal cycles and between years. Furthermore, only some portion of terrestrial precipitation going into runoff in rivers and into groundwater recharge can actually be used, for technological reasons as well as because of environmental conditions (regional and seasonal differences in abundance, in particular) and should be used because of the dependence of ecosystems on sufficient amounts of water.

The very beginnings of civilization are related to the utilization and management of water, but only since the advent of the fossil-fuel economy in the last 50 years – with its capability for widespread water retention by dams, and diversion contrary to gravity-fed courses over large distances by pumping technology – has water management been focusing all but exclusively on supply-side infrastructure approaches: more than half of the world's major rivers have been modified by dams and diversions (UNEP 2002/WCD 2000), with water being taken from several of them to such a large extent that they do not reach their earlier end points any more, e.g. the Amu Darya in Central Asia which feeds (or rather: would feed) the Aral Sea (Brown 2001).

Some 70 percent of the diverted and pumped water worldwide is used for irrigation in agriculture, and this demand is expected to increase as world population increases, even while rising industrialization and urbanization require that more water be diverted to industrial and household utilization. The deadbeat argument that an increase of irrigation in agriculture will be necessary to provide enough food for the growing world population, however, is rather dubious. As the case of the Amu Darya attests, the actual process can be that the irrigation water goes towards growing cash crops (in this case cotton) rather than to producing food for people (and it resulted in the loss of the economy associated with fishing, which provided necessary protein and work). Moreover, possibilities for changing (subsistence) agriculture towards systems utilizing traditional, locally fitting and/or new highly efficient technologies, and species and varieties of crops adapted to local conditions, are all but completely unexplored, and commonly being lost rapidly where they would still be in existence because they are socially and economically undervalued in relation to industrialized agriculture and products.

Furthermore, increases in water supply to human use have been achieved by pumping water from deep (fossil) aquifers, which are virtually non-renewed/renewable, and shallow groundwater, which is renewed at lower rates than those with which it is currently being pumped out, as shown by falling water tables. Despite some changes in water policy towards an increased focus on changing demand (UNEP 2002), most current plans for solving the problem of increasing water scarcity still attempt to institute supply-side "solutions" utilizing what Josephson (2002) calls "brute-force technologies," e.g. dams, diversions, deeper wells. These, however, simply cannot be sustained in the longer term, e.g. as aquifers become depleted (and in coastal areas even experience an influx of saltwater as a result; cp. Postel 1997); and the longer such an increasing, unsustainable diversion of water from little or non-renewed sources and out of natural ecosystems continues, the more disruptive the ecological changes wrought and the social changes that will eventually be imposed.

Additional concerns arise from the high requirement of technological and specialized-labor inputs these large-scale programs of water transfer require, which may prove impossible to uphold if social disturbances were to arise – and conflict over water is very likely to become a major concern in the future and has been starting to be already, both regionally (with questions of allocation to different users), inter-regionally (e.g. with schemes for diversion in Spain, cp. Chatel & Steinweg 2002), and internationally, although it has yet, by and large, been resolved through diplomatic efforts. Additionally, such constructions present themselves as ideal targets for terrorist attacks (unfortunately, a major concern in recent times), which would affect a large number of people directly and indirectly in a short time (which they tend to do already e.g. in the case of hydroelectric dams when extraordinary climatic events decrease water flow – and incidentally even with nuclear power plants when river water used for cooling gets too warm – or topple power lines and thereby cut people off from the provision of electric power, and earlier during construction if expropriation and resettlement of local people is required). Moreover, the focus on growth, e.g. of supply of water, and economic value derived from it, fails to take the wider ecological-systemic embeddedness of the issue, as well as eco-cultural dynamics, into account.

Ecological Interrelations

Water is necessary for all life, as well as essential for the functioning of ecosystems (and the defining feature of aquatic ones, of course) which provide a host of services and natural capital, including an influence on the distribution and quality of the water itself.

First of all, ecosystems at catchment areas influence the usual (assumed to be "natural," i.e. not contingent on biotic-ecological factors, taken for granted) patterns of precipitation, notably in the Amazon's recycling of a large proportion of its water (half to four-fifths according to Salati & Nobre 1992), and illustrated by changes in rainfall attributed to deforestation (Meher-Homji 1992). Together with glaciated areas and groundwater, catchment ecosystems in general mediate the distribution of water over time, e.g. (rain)forests in the humid tropics which "exert a 'sponge effect' and soak up moisture before releasing it at regular rates" including during the dry season (Myers 1997: 216). Old-growth coniferous forests in watersheds of the American Pacific Northwest, for example, exhibit such an effect as well – due to their extensive volume and crown surface, they effectively comb water moisture from clouds and fog (Harr 1982, quoted in Franklin 1988: 168). Furthermore, by the associated effect of dispersing the force of downpours, storing water, and also by holding soil, catchment vegetation protects against mudslides and avalanches, and together with floodplains and wetlands, mitigates fluctuations in water supply. Less drastic but in the longer term just as deleterious forms of erosion, which partially constitute a loss of fertile topsoil and lead to siltation of reservoirs, are heavily affected by different forms – or the absence – of land cover, as well.

Moreover, water quality is improved in its passage through ecosystems, and, as the well known example of the Catskill watershed – the water source of New York City – shows, this service can be a highly cost-effective alternative to the building of filtration plants even when only water quality is the consideration: the necessary water filtration plant was estimated to cost 6–8 billion dollars for construction and 300 million dollars annually for operation, the alternative watershed protection cost an estimated 1.5 billion (Daily 2000). In the humid tropics, access to more, cleaner water would most easily be achieved by assuring that watershed forests can perform their natural ecosystem service, but the lesson from the Catskills, in contrast to the economist's "get rich now, clean up later", has not been received: "public-health programs in [various] conurbations ... are being set back through deforestation-caused declines in quantity and quality of water supplies" (Myers 1997: 219).

Freshwater ecosystems provide further services and capital (cf. Postel & Carpenter 1997, Ewel 1997), which benefit humanity in manifold ways, most directly by extractive use through the provision of fish, shellfish, waterfowl, and other (e.g. fur-bearing) animals, and wood, fiber, and peat.

Related to this utilitarian value is the provision of wildlife habitat, a precondition for such resources, as well as an "existence value" and ecosystemic benefit itself. As habitat, freshwater ecosystems (including marshlands that may contain brackish water rather than actual freshwater) harbor specialized or even endemic species that are of particular value in terms of biodiversity, but are or may be useful in ecosystemic terms, e.g. of water cleansing and by their contribution to global biogeochemical cycles, and as genetic resources for new or improved agriculturally useful plants. Even in more narrowly economic terms, many of these ecosystems are of high value as breeding/spawning grounds for freshwater and even some marine (salmon being the prominent example) species used by extraction.

Associated "para-extractive" use, where the activity rather than the extracted good is the major "product," i.e. recreational hunting, fishing, and (less extractive still) birdwatching, can be another major source of revenue from aquatic ecosystems which is dependent on their ability of serving as habitat.

Non-extractive/in-stream benefits, e.g. transportation, hydroelectric generation, and dilution of pollutants, similarly depend on a flow of water of sufficient size; hygiene (washing, bathing) may not be an in-stream use anymore, but used to be – notes on sanitation will follow.

Socioculturally-contingent benefits include the utilitarian value of aquatic systems for recreation (and the enhanced value of property at the waterfront – though the opposite can be true as well if inundations occur). Moreover, landscapes with or bodies of water can not only provide for vacation spots with spectacular and/or soothing views, oftentimes they are even accorded spiritual value, e.g. as sacred places, mythological-religious sources of absolution and founts of life. Moreover, even where sacredness of the actual feature is not the issue, it may still suggest metaphoric/symbolic value in addition to the utilitarian, or serve as the locus of/for traditional environmental knowledge (as any dedicated fly-fisher would probably tell you, these issues need not be quite so separate in personal approaches to valuation).

Agricultural, municipal, and industrial usage of water requires its diversion while the host of other benefits described above depends on sufficient water remaining in ecosystems. Were benefits from pollution dilution (and processing), in-stream usage, and provision of habitat set against the profit from agricultural production with irrigation, less diversion for this later purpose would be justified (Postel & Carpenter 1997); but diversion for industrial utilization which currently provides much higher economic value than other uses – but ultimately is less fundamental in terms of sustainability-as-survival, and may not provide higher value compared to the aggregate of in-stream uses – may be another matter.

Changing Course, Not Streams

For the consideration of sustainability in the "positive ecology" outlook, it is necessary to go somewhat beyond the discussion of the need for water to the natural capital and services it provides in and with (aquatic, watershed, etc.) ecosystems. These do show that the common economic valuation of water as a cheap resource to be channeled – actually at great cost to society – wherever society supposedly wants it is myopic, particularly in those cases where water is "supplied for waste," i.e. used excessively, inefficiently, and in ways that leave it polluted, and sometimes even "supplied as waste," carrying harmful microorganisms and chemicals because of insufficient watercourse protection (from influx) and treatment. Particularly salient in this regard is the continuing, conscious and accidental, approach to considering pollutants washed away to be gone when they have an unfortunate tendency of accumulating in food-webs and returning back through this cycle.

Highly efficient distribution of water drawn from little or non-renewed sources, similarly, qualifies as what Watzlawick (1994 [1986]) called a "pat-end-solution", a solution so effective it ultimately gets rid of the source of the problem it was instated to resolve in the first place (such as the people in need of water).

The challenge of providing a guide to action, however, requires that the alternative possibilities for a use of water be shown: A use that enables both survival and improved quality of life by providing, as the South African water policy (reported in Hawken, Lovins & Lovins 1999) commits to, "some – for all – forever" while utilizing only a portion of total renewable freshwater supply available locally, avoiding both large diversions (so as not to interfere with ecosystemic functioning and provision of services/natural capital), and the usual focus on large-scale "solutions" (for the disturbances and insecurity they represent). Indeed, many of the puzzle pieces for a more sustainable use of water do appear to exist; views into the past and at recent innovative approaches come together in them.

After all, the widespread utilization of freshwater resources that are not naturally renewed, combined with a wide abandonment of any techniques for working with renewable water supply in elegant – simple but highly efficient – forms is a rather recent development. Such "harvesting" and management of rainwater had been working for centuries, however, e.g. in the Andes (among the Inca), in Africa (among the Chagga), or in India, where it may have evolved as response to climatic changes and could hold lessons for promoting societal resilience in water-climate-agriculture relationships towards the future, again (Pandey, Gupta, and Anderson 2003); and could now be combined with methods for water quality improvement and even more efficient usage, combining:

1. Techniques of rainwater harvesting with cisterns, check dams (small structures adapted to local topography to contain water), agricultural utilization in traditional ways – particularly as far as using and (re-) discovering/developing adapted plant species, varieties, and (e.g. diverse, resilient) agricultural systems is concerned – and in modern highly efficient ways, e.g. drip irrigation.
2. Techniques of water quality improvement: utilizing the ecosystem service, possibly even in ("eco-") industrial settings as with "Living Machines" (described in Hawken, Lovins & Lovins 1999). Further disinfection of "ecosystemically cleaned" water can then be achieved by relatively simple, cost-effective technology , e.g. the Miox system (cp. www.miox.com, usable even with solar power), or even directly using the sun (cp. www.sodis.ch). Even the occurrence of cholera as a water-borne disease can be reduced by simple filtration with sari cloth (Colwell et al. 2003). Still, the necessity of prevention is paramount: sanitization of water cleaned through ecosystems functions and not containing persistent pollutants is relatively easy to achieve by technological means, completely technological cleaning and sanitization, as well as treatment to remove particular pollutants, is costly and complicated.
3. Multiply utilization, such as with "brown water," water that had been used in household situations like showering, which could still be used for toilets, at least, but better still for passing through/into (created) freshwater ecosystems and possibly (if dangerous pollutants are kept out) into agriculture.

Sanitization, of course, does need water, but can work effectively with relatively little of it; as far as human excrement is concerned, water should not be used whenever possible (let alone drinking quality tap water in large amounts, as is still very common). Rather, "dry" approaches should be applied, either by re-separation or better still with composting toilets – instead of producing water pollution, they prevent one of the "two biggest components of the global water crisis ... the contamination of drinking water supplies with human faeces" (Nature, 20 March 2003; the other being the wasteful application in agriculture) while contributing to, well, the closing of nutrient cycles between human beings, food, and land (cp. Esrey & Anderson 2001, "Ecological Sanitation. Closing the Loop;" Esrey et al. 1998).

The main requirement for bringing these approaches to fruition, however, is a scientific-inventive (not necessarily expert, but widespread) attitude to the challenges presented by the relation between the need for water and the ecological underpinnings, starting out from the requirements for sustainability of "fitting in" with and supporting ecological functioning (for example including groundwater recharge for gaining a backup of supply and, together with sufficient river flow, ensuring ecosystem functioning), helped along by political will. Unfortunately, the necessary shift of focus from the "monumental solution" to achieve great change in a single step, to locally (ecologically/environmentally and socially) adapted solutions appears to be the greatest and most difficult of steps, especially as long as it is – discounting "externalities" (such as pollution, ecosystem change) and longer time-frames – seen as counteracting economic growth. Like all the regards of sustainability, failure to consider them now will not, however, make them go away, but only threaten security, and with it development, in the future – if not already.

Chapter 11

Nutrition

The human need to eat, satisfied mainly by providing food through the creation of agroecosystems, is one of the concerns most obviously and widely connecting sustainability with the issue of human population size and growth, with the human relationship to (other) structures and functions of the ecosphere, as well as with further relations between individual choice and experience/enjoyment, and with culture(s).

For its importance in these various – essential, utilitarian, as well as cultural, emotional, and experiential – aspects, and its deep interrelation with ecological processes, nutrition is likely to be just as pivotal an issue for the future as it has been for humanity's history to date.

At the same time, however, it is an extraordinarily complicated issue, as political-economic, technological, cultural, and ecological influences all bear on it. Industrial, "green revolution" agriculture is commonly presented as the inevitable progress in food (or rather renewable resource) production, with further intensification needed to feed the growing population. Therefore, there supposedly were no need to consider an alternative to it, nor could there even be one. As a matter of fact, considering the foci of sustainability and their requirements, however, there are obvious reasons for holding industrial agriculture to be particularly problematic, even while the potential of sustainable agriculture (and especially scientific validation of it) is only just coming to the fore – although, in historical perspective, such "alternative" forms of agriculture have sustained humanity for a much longer time than the fairly recent industrialized ones.

Then again, it must be clear that human nutrition, both concerning agroecosystems and their balance between production and conservation, and diets and their balance between scarcity and overeating, is about a middle path between extremes; there has never been an Eden where agriculture was not a gamble between human and natural processes of positive and negative effects – crop development, yields, and resilience; agricultural techniques; crop failure, drought and flooding, pests, and so on. Yet, there is reason to hope that true agro-ecosystems could be developed – after all, some historical examples of biodiverse cultural landscapes do exist – that can be sustainable ecologically, fitting in with and even as somewhat wild habitats, promoting their own productivity and resilience, as well as providing the amount and diversity of food necessary for human well-being.

Contexts and Misconceptions

The usual center of attention in discussions concerning food and agriculture, at least ever since Malthus, has been on the prospect for feeding an already large population that is growing still further. The apparent success of industrial agriculture in raising crop production is taken as indication of its rightness; that hunger still occurs typically attributed to a lack of food supposed to be occurring nonetheless. In spite of the voices arguing that it were inevitably necessary to industrially intensify agriculture further to increase food production, scarcity appears not to actually be the reason and result of the current state of world agriculture.

For one, the mere numbers e.g. of world crop harvests speak otherwise: There would be sufficient agricultural production, not only if less area were used for cash crops but from current area, to feed current human population of the earth, and even enough to feed a much larger human population if used directly as food. A large proportion of crops is being fed to animals for meat production, however (a relationship Frances Moore Lappé already pointed out in 1971, and that still holds true).

The need for increasing intensive production is even more questionable considering the other simple ratio that a similar number of people now suffer not from lack of food, but from overabundance, obesity and the health problems associated with this "epidemic" (Nestle 2003, Gardner & Halweil 2000). This relation not only holds true across global statistics, but apparently, even within the same country, the USA for example, both forms of malnutrition coexist (cp. Schwartz-Nobel 2002), and even exacerbate each other as poorer segments of population can only afford less healthy diets, contributing simultaneously to a lack in micronutrients and an excessive intake of sugars and fats.

As Lappé et al. (1998) argue in their analysis "World Hunger," most of the views treated even as articles of faith may on closer inspection turn out to be myths, most importantly that of scarcity (of food resources vis-à-vis the world population) caused by insufficient production and losses in the struggle against nature. The latter relation stands in even starker contrast considering that attempts at controlling the conditions of food production, e.g. pests and nutrient availability, by technological-chemical means result in negative side-effects such as pests becoming resistant, chemical input changing the dynamics of the agroecosystem and even global cycles, while agrotechniques working with ecological processes can obtain similar results with positive synergies (see below).

Two issues in the context of sustainability and food production are particularly interesting. These are the mutuality of the relation between food and population, and the misleading simplicity of supposed solutions to the various problems, which misunderstands the actual complexity of the relationships involved.

Food Availability and Population Dynamics

The negation of any realization that food availability would, directly and in conjunction with cultural, psychological, etc. factors, limit or support population growth and size is telling. The existence of some such relationship is straightforward from an ecological perspective, even if human creativity can shift the numbers of carrying capacities making the concept little useful in the human context. The problem is that it is usually negated in full, i.e. population numbers are treated as an independent variable, while the food necessary to sustain them is seen as the only dependent variable. – From this limited perspective, human population growth is at once equated with a need to increase food production. Yet, the relationship is existent and mutual (cf. Hopfenberg and Pimentel, 2001 – who are about the only recent authors to mention this dynamics).

So, only arguing for increased food production will not stabilize population dynamics, as increased food availability could not only support but be a causal factor of yet further population growth. Therefore, the issue of population would need to be addressed directly, as well, if sustainable numbers and relations are to be reached. As noted before, – mentioned in discussing population as a fundamental context – there are indications that a sense of security in life that does not require large fertility to ensure survival (of children, and/or of the parents at old age) achieved with education and medicinal infrastructure, for example, may be highly influential (if not sufficient) to alter these dynamics in a humane way. Current trends related mainly to economic, and in some cases political, developments appear to be moving human population size towards a stabilization at the end of the next half-century. At the high number it is now suggested to reach by then, it seems likely that "the world can be fed," and be inclined not to grow further because of factors other than food security.

Cycles of further expanding population growth and – and actually caused by – agricultural intensification would in the absence of such stabilizing tendencies yet have the potential to continue until they have reached their ultimate limit and population numbers start to be reduced by "natural" means (cp. Hopfenberg & Pimentel 2001). So far, the prognosis of Malthus has fortunately not materialized, and it probably is not going to, but unfortunately this does not mean that it never could, at least in confined areas, nor that a larger-than-ever human population would not run into other problems. – A mainly agricultural and urban, biologically impoverished planet may still be capable of supporting humanity physically (cp. Jenkins 2003), yet where ecosystemic thresholds effecting a change to new states less amenable to human existence lie at a global level is all but known, and even if these were nonexistent, this does not mean that such a state of the world would not adversely affect us on other (e.g. psychological) levels, and betray human moral capacity in the loss of diversity and potential it entailed.

The Simplicism of Solutions

"Solutions" to malnutrition, too, just as simple-mindedly fail to acknowledge interactions, arguing only for raising productivity if quantity is the concern, or, if the problem lies with the composition of available food (content of macro- and micronutrients), engineering single crops so that they produce the one nutrient the diet among the population under consideration is most deficient in. Further, especially cultural and ecological relationships that are always of influence are not taken into consideration.

Thus, the linkages between globally necessary food production, usual and preferable human diets, and health/food security largely go unnoticed, so that the additional benefits more comprehensive approaches would provide fail to be acknowledged. The change in human diets that accompanies "development," for example, towards increased consumption in general, and especially of fats (meat) and sugars (but not from fruits, and not of vegetables) is seen as a "natural" process. Therefore, there are few attempts at resisting these trends. Promoting diets less conducive to obesity will become necessary, however, since they are preferable considering both population health and global food security (e.g. since production of meat in intensive systems for mass consumption utilizes about seven times the crops for one unit of meat, while more direct human crop consumption and less consumption of meat enables the production available by letting ruminants graze, turning otherwise not human-usable grass into usable meat without need of crops, to suffice; approaches like these may provide quite enough output for a diet that is quantitatively sufficient as well as healthfully balanced).

Yet, the single problem/single solution-approach to malnutrition is already questionable at the very least considering that a healthy diet is not one that tries to provide the one most lacking nutrient, and then the next most lacking, etc., one after another, but rather a good balance and/or high diversity that is likely to provide an ample range of all essential macro- and micronutrients. The positive effects of diverse diets are rather well documented, in fact, especially in the ethnobotanical literature on dietary composition and use of noncultigens, even suggesting the occurrence of additional preventive and therapeutic effects (cf. Etkin 1994). Moreover, it has been argued for the case of "Golden Rice" as means against vitamin A deficiency, that socio-economic and environmental issues were actually the major factor, while its solution by supplements (of which this rice would only be another one) has consistently failed (Egana 2003).

In the same vein, biotechnologically engineering plants to adapt them to different environments is – because of its appeal as high-tech "silver bullet"-solution – seen as an element of progress, while species, varieties and landraces already adapted through local cultivation are being lost for lack of support by "progressive" farmers and markets.

In general, examples of agriculture that is more likely sustainable, in contrast to techno-agricultural utopias, are commonly outright dismissed for similar "reasons," arguing for example that output were not high enough (without considering any other factors or even actual numbers, for that) or that "organic farming ... is no more sustainable than the fish-farming that produces high-value smoked salmon to those consumers who can afford it" (Trewavas 2002: 670), as if current prices which are strongly influenced by market distortions such as subsidies were an adequate indicator. Rather than any one, single factor, the conjunction of ecology, agriculture, and culture (including human needs and economic patterns) needs to be considered.

Efficiency? – Economics and Food Security

Industrial agriculture's problems which make it incapable both of being a solution to hunger and of being sustainable can be made increasingly clear by the consideration of eco-(agri-)cultural conjunctures, although the same multifactorialness does make for a nearly intractable matter.

Its supposed success, for example, has been achieved by trading in some of the dependence on environmental conditions for an (additional) "techno-environmental" dependence on chemical and energy input (from fertilizers to pesticides), most of which derives from fossil resources which are ultimately limited. Moreover, the maintenance of this intensive system by human input is actually more of a difficulty than achievement of its viability by ecological processes would be, particularly as it interferes with many of these ecological processes (that have been making it possible in the first place), so that its success is necessarily limited to the short-term (cf. Matson et al. 1997, Jordan 2002).

In effect, a system of using plants' (and indirectly animals') ability of transforming solar energy, with the help of water and nutrients and an expenditure of energy in the form of guiding human labor, into chemical energy usable to humans (a.k.a. food) has been transformed into a way of producing (and transporting over long distances) any calorie of energy by using the same to several times that number of calories of fossil energy (Odum 1971, Hall et al. 1992). In contrary to the usual self-promotion of industrial agriculture, this relationship disproves the supposed efficiency of this system, justifying the verdict of Thu and Durrenberger (1998: 2) that, "(b)ased on measures of energy expended per calorie of food produced, industrial agriculture is the most inefficient form of food production in the history and prehistory of humankind."

Even considering more immediately economic and developmental concerns, farmers have been made to be dependent on the industry providing the machines and chemicals they need, and even the seeds they plant, bank loans to pay for those inputs, and government subsidies to make this system of wrong incentives continue (cp. Myers & Kent 2001 for the latter). Except for the comparatively well-off, especially in developing countries, this is a common route into economic trouble or even poverty, as soon as purported high-yield crops fail because they did not grow in the climatic conditions, or did not receive all the input they needed (e.g. for lack of funds to afford it). As a consequence, peasants and farmers tend to lose (or give up) their land (cp. Boucher 1999).

A misrelation between actual farm operation, agricultural production, and "supporting" industries is problematic even in industrialized countries, where farming produces only about 1 percent of GDP, but the entire farming-related industry and trade fourteen times as much (Hawken, Lovins & Lovins 1999: 191). Increasing farm output is a very short-term solution to keeping individual income at former levels, but the higher productivity does, and has, not surprisingly lowered farmer's real income (ibid.).

On the other end of the line, where the harvest enters the market, additional distortions provide a problematic impetus, as subsidies (e.g. for supporting the export of the produce of farmers in industrialized countries), and pressure of trading companies (demanding ever lower prices), for example, decrease the price of farm products further.

The composite effect not only makes farmers' survival more difficult, but affects the nutritional situation in both developed and developing countries:

In the former as the overconsumption of cheap, abundant, heavily marketed foods, even cheaper in the production (having increased it "to feed the world"?) and of low value in terms of what would constitute a healthy diet, but suiting innate human food likes, lets profits of the food industry grow, together with the population's waistlines and public health costs (cf. Brownell & Horgen 2004).

In further-away places, the situation is similarly problematic. In some cases (cf. Ratta & Nasr 1996 quoted in Bryld 2003), the "structural adjustment" in food production has led to an increase in prices, cutting the poor segment of population off from food supply. This mechanism, as described by Ratta and Nasr, operates in at least three ways, through price increase, devaluation of real wages, as well as loss of land supposedly inefficiently cultivated. The latter includes the cultivation of traditional food crops which provide yields high enough for subsistence, but which are not traded widely or not even marketable (only a very limited number of crop species and varieties is; see below) – and therefore not counted into GDP (?).

Norberg-Hodge (2000), providing a contrary outset with the same results, reports that imported food stuffs are now cheaper in Ladakh than locally-grown ones, moving people away from (subsistence) farming. Other work yet remains harder to come by – and in contrast to the small percentage of farmers in "developed" countries, (subsistence) farming is a major economic sector also in terms of people employed in it in "developing" countries. At the same time, even traditional products and ways of production that would be considered progressive from the sustainability and even the "Northerner's" point of view come to appear outdated.

Such developments are among the little acknowledged tragedies affecting the global nutritional situation, cash crop-oriented and other modern agriculture, and food aid alike (cf. Boucher 1999): feeding the world rather than helping it to feed itself may be producing learned helplessness all by itself where the produce ends up, the producers see little they could do but to increase output (even if the combined effect is decreasing real income), and this situation is further aggravated by the promotion of industrialized, unsustainable agricultural systems. The latter's "advantage" seems to be shown by the food prices, by making locally produced foodstuffs compete (unfavorably) with heavily subsidized, unsustainably produced ones, by the marketing for high-yield varieties and the products made from them and heavily marketed to keep making a profit while food prices decline by promoting (increasing and over-) consumption. Local, even potentially sustainable agroecosystems are meanwhile converted (both out of seeming necessity and because this move towards resembling industrialized countries' economies is being promoted as progressive) into food or, more usually cash crop, monocultures the profit from which is to pay for existing debts as well as for those foods that now have to be imported (or seem to add value rather than cost in developed countries by being transported, and of course processed).

Yet, in cases of need, poor people do not exert a market pressure, and would therefore go hungry if they do not (or cannot even try to) provide for themselves – and most of the time, the change to industrial agriculture goes hand in hand with poorer population's loss of land (and the inheritance of colonialism already is a concentration of land in the hands of a small elite, cf. Lappé & Collins in Boucher 1999, for example), so that they come to be jobless or in need of jobs which are particularly difficult to find for this segment of the population.

In this context, people's education plays a large role in two ways, one, as the poor – often farmers – tend not to be highly educated and could therefore only either farm or engage in simple forms of manual labor, in factories, providing services on the street, etc. – Yet, secondly, sustainable forms of agriculture needed to constitute a knowledge economy, combining just such rather unskilled manual labor with the deep agroecological knowledge required for them (as will be seen in the course of this chapter). Potentially, acknowledgment of the latter as the progressive agriculture it is would therefore offer a way to improve the standing of smaller farmers who have been practicing something akin to organic agriculture because it has been the only viable and/or available practice (cp. Parrott & Marsden 2002), and of sustainable farming in general.

Expanding the agroeconomic discussion further and more directly into trade relations, additional dynamics less commonly considered should be addressed:

Seen in a strictly ecological/ecoregionalism-based perspective, the relation between food production possible in the local area and population (able to be) supported by it would, at least ideally, need to provide the guideline for sustainability.

Trade is an excellent way of supplementing locally available resources with products that cannot be grown in the region under consideration, and to receive luxury items – as it has been since earliest human history. Locally unavailable staple foods can just as well be imported, but it is here that the above, eco-regional argument comes into play: A division of labor where tropical countries grow tropical produce, regions with deep topsoil concentrate on staples, etc. does present economic (and ecological) advantages; but trade in the most basic foodstuffs as opposed to their local production and consumption is likely to pay off only as long as high subsidies and low labor and transport costs make it viable, but not if the market, particularly counting externalities in, really worked as it is theoretically supposed to.

Moreover, the alternative of import is a source of insecurity not just to the poor population not exerting enough market pressure, but even to and within industrialized countries, e.g. as most urban centers would be hard hit if terrorism or oil shortages restricted transportation of food (Halweil 2002).

Additional problems arise from the increasing development of such a division of labor where most areas/countries try to move into economic sectors seen as providing more value than agriculture, relying on economic clout that they hope to achieve for the provision of food. However, food import is not an only approach viable in the long term if employed well near globally, leaving food production only to either the poor/unskilled, and/or to factory farms in which ever more input takes the place of ecosystem services, although it cannot do so in the long run.

Imagining, for example using the case in point Brown (2001) puts forward, that China gave up its orientation on self-sufficiency and focused on technological-sector economic growth, and managed to grow rich enough with it, this single country's demand for agricultural produce could greatly change global food prices (even so, a similar pattern may already be emerging, cp. EPI 2004), adding volatility to the insecurity inherent in such a system of reduced redundancy and (in result) less resilience. At the same time, most other countries are attempting to follow such a pattern of development, but it encounters the problem with material standard of living/consumption that the "human footprint" presents, i.e. that it would require more than (the resources of) three planets Earth to be possible, and great increases in agricultural production in those few areas of the one Earth that would then truly be responsible for feeding the world. The unsustainability of "modern" forms of agriculture adds to the problems.

Moreover, there are other social effects such as the loss of jobs "achieved" through the change from smaller-scale farming to large-scale industrial farming and other hopeful, but probably labor-reducing, kinds of business. These affect both "developing" and "developed" countries, and are even supported by government policy in both, as low labor and high technology are considered the only good business practice; on the other hand, neither numbers and diversity of employment, nor values of community, quality of life, etc. that could be upheld or achieved by moving away (or not even into) industrial agriculture are considered (Thirsk 1997, Daniels & Bowers 1997).

Bringing together the considerations of economics, employment, and the agricultural sector's situation, the greatest challenge will be to find a balance between the remunerative potential of employment in other economic sectors – so that education and improved infrastructure does offer a way out of poverty by getting jobs there, if possible – with the impossibility of the whole world following this path of "development," and the ultimate importance and value of food production, combined with the necessity of a change towards sustainable forms of agriculture, which suggests both that a higher economic importance and value of the agricultural sector (but without external input that channels the income straight away from the farmers again, and complicated by related factors, cp. Strange 1988 [1999]), and a higher standing of farmers in a new kind of knowledge society may become a requirement.

It is comparatively easy to analyze the differences between different systems of agriculture, and to contrast dominant and alternative ones, at least.

Regarding their efficiency and yields, which have been the basic arguments for industrial and against organic agriculture, for example, there are increasing suggestions of the viability of the latter, not least with a recent Science article (Rosegrant & Cline 2003) mentioning "agroecological approaches offer[ing] some promise for improving yields:" Input of (fossil) energy and fertilizer is 19 percent and 30–60 percent lower, respectively, under biological agriculture. Therefore, even if yields were lower, efficiency would be higher (Maeder et al. 2002). Furthermore, with food being bought in the region as opposed to being transported over long distances, four to 17 times less fossil energy is used (and with less dependence on extraneous input and transport system, food security is improved at the same time, cp. Halweil 2002).

Moreover, organic systems can provide yields as high or, particularly under suboptimal conditions, higher than those of conventional agriculture (Rodale Institute 2000), even while requiring much less external input but only different management of the various components that make up the agricultural system, e.g. in the "system of rice intensification" (cf. Uphoff 2003).

Additionally, the diversity of crops, both of varieties within cultivars and between species, the variety and flexibility of agricultural systems, as well as their coexistence with wild areas providing additional foodstuffs, e.g. greens, herbs, and famine foods (cp. Huss-Ashmore & Johnston 1994), as found in alternative/traditional agriculture provides a resilience and safeguards unknown to the simplifying rationality of currently conventional agriculture (cf. Nazarea 1998; Thirsk 1997 for a historical perspective).

By way of example, consider the situation in the Vietnamese-Chinese borderland (the nuclear area for domesticated rice), where "some highland villagers will grow as many as ten varieties of rice, each suited to certain environmental conditions, so that if the weather is bad or the crops are infested with pests, the likelihood is higher that at least some varieties will thrive" (Proschan 2003: 66); moreover, such diversity can produce an effect known as overyielding, wherein a number of species/varieties growing together produces more yield than any of them would by itself (Hooper & Dukes 2004).

Ultimately, (agro-) ecosystems can thus be "both productive and protective" in providing habitat and ecosystem services (Jordan 2002: 155).

Ecological Interrelations

Considering the continuing growth of the world's population, even if conversion to agricultural land or increased productivity achieved by utilizing inordinate amounts of material and energy input cannot continue forever, both further land conversion and (or?) ongoing intensification seem to be necessary in the short run. It may appear that loss of biodiversity, ecosystems and their services is a natural consequence. In the longer term, however, the viability of such a system eventually running out of land as well as destroying the very capital of ecosystemic processes it is based on is more than limited; more of the same only shifts the onset of problems a little way back, but does not eliminate it.

Actually, it is estimated that all land well suited to agriculture is under cultivation already, so that its better, sustainable utilization is the most promising way forward both for food production and for ecological conservation, as expansion into additional areas would not change the state of food supply significantly, but affect ecosystem services (and in logical conclusion, ultimately agroecosystems) negatively.

"Eco-agriculture," focusing on its own sustainability as well as integration into the wider matrix of the land, on the other hand, may be capable of providing both enough food yield and security, and ecological conservation of remaining natural, less human-dominated areas, as argued by McNeely and Scherr (2001), even in and with tropical forests as in agroforestry (cp. Atran 1993, for example). The diversity and potential of "integrated farming systems" in combining different features, working at different scales in different situations can be found in China in its entire range (cf. Li 2001) – from phyto-animal systems to agro-silviculture/agroforestry, multiuse-forest, poly-aquaculture, and the use of rice paddies as fish and fowl habitat, from homestead gardens to regional approaches, and from arid lands to wetland ecosystems. With the potential of agroecological sustainability as implying a co-existence and utilization of "wild" elements within what is then truly an agroecosystem, there even is considerable possibility that farmland itself could provide "natural" habitat (cp. Jackson & Jackson 2002).

Environmental effects which represent the second major way (besides dependence on inputs) by which currently modern agriculture is made to be unsustainable are now only just coming to be considered in depth. They extend from limits to land conversion to agricultural use, to loss of soil and soil fertility, and to direct and indirect impacts of agriculture on ecosphere function, and vice versa; and they include relations with crop and wild biodiversity, as well as concerns of human life; effects which do not find their way into dominant, "rational" considerations of cost-benefit and efficiency. Once such externalities start to be rightly considered, the problems and costs of industrial agriculture become even more obvious, in strictly monetary, (macro)economic terms as well as in terms of other relationships and values, ecological and social alike.

Rather than discussing only these problematic relationships still further, however, it is time to consider ecological and cultural relations in the context of the case(s) for sustainability in agriculture, which also serves to put these less often considered problems into perspective. Therefore, some of the major problems affecting agricultural sustainability will have to be mentioned, at least to serve as counterpoints, but not gone into in detail as they have been presented in depth elsewhere (e.g. Kimbrell 2002).

The major question, after all, is yet again whether and/or in what ways the need for nutrition could be addressed so as to support human well-being as well as ecological sustainability, in alternatives conjoining ecology, agriculture, and culture. Both historical cases for (cf. the discussion of "ecosystem people"-perspectives) and contemporary developments towards a possible "integration" between human needs and ecosystemic conditions attest to the potential of an agriculture less destructive of wild places, and conservative or even restorative in regard to its own ecosystemic foundations and relations.

Should such approaches be capable of exerting enough pull to achieve wider change, it would be all the better, but the political-economic changes necessary to reduce or remove market distortions will probably need to be taken on in those arenas. At least as far as distortions by subsidies on production size and exports are concerned, they are finally made an issue in trade talks (e.g. the – failed – WTO talks in Cancun in 2003) and recent EU agricultural policy. Even a market less distorted by these remains little responsive to the costs presented by ecological "externalities," however.

The decisions for or against certain forms of agriculture are a cultural, attitudinal problem to a large extent (cp. Morris & Andrews, Duram, Gilg & Battershill, Curry-Roper, all in Ilbery, Chiotti, and Rickard 1997); of course, the edge of lower cost and higher achievable prices with sustainable practices can be a factor as well.

The commonly higher requirement of sustainable agriculture for labor, for example, naturally relates to farmer's and potential farm worker's attitude and like or dislike for hard work. Additionally, however, it is a problem of national politics – taxation is a major factor influencing whether more labor is a viable option; the income from agricultural production, of course, is another major issue bearing on this factor, and related to politics as well as to economic policy.

Still, knowledge of the very existence and progressive nature of "alternatives" (which in traditional, mainly subsistence agriculture have been, and in part still are, actually the dominant forms of agriculture), putting them into relation with sustainability's dual concerns, with their meaning to both society and the individual, and import to modern and future conditions, is the necessary first step.

The most immediate relationship between humans and (agricultural) land that comes to mind in the context of sustainability would be the just-mentioned necessity of land conversion from the wild state to the cultivated state, most prominently drawing the connections with ecology and conservation, and with basic human needs as well as social issues, respectively. Apparently, some change and/or the change of some areas for agricultural use could be regarded the (almost exclusively) human, natural way of providing for nutrition; natural, just like other animals' behavior of not only finding, but to some extent actively (albeit less planned, of course) constructing their ecological niches. The verdict that most land well suited for agriculture were already under cultivation, however, limits the extent to which further conversion would be possible; sustainable agriculture, then, may need to go hand in hand with intensification of agriculture on lands presently used for it, under the condition that novel ways based on and supporting ecological functioning are put into practice.

In quite a few areas around the world, landscapes that are now considered natural by most people and that are ecologically functioning and biodiverse have in fact been created by agriculture; others could be conserved in spite of or with cultivation. In some instances, and very prominently with industrialized agriculture, however, agriculture had and has been working against the environmental circumstances that it operates under.

Both possibilities can be seen in cases of deforestation for agriculture. In Europe, for example, many areas that used to be forested have been brought under agricultural cultivation during the course of millennia, and are now prized, culturally and ecologically valuable landscapes of the Mediterranean (where deforestation also occurred for shipbuilding and other such uses) or in the Alps. Shifting/swidden agriculture in tropical forests, too, appears to have continued without overwhelming negative effect on the natural ecosystem and its diversity while population sizes were low and/or forest areas large enough to allow for sufficiently long fallow/recovery times; but currently, with other approaches than in historic times and in situations where these preconditions are (no longer) extant, these agricultural systems fail.

The most fundamental, commonly less immediately considered relationship of cultivation and land that these different conditions ultimately hinge on is the dependence of agriculture on soil and its fertility. Therefore, this point forms a strong focus for a sustainable agriculture that is viable in the long term and need not expand the size of areas under cultivation.

Further ecological relations arise from the connections within agroecosystems and between them and wild ecosystems, and with crop and wild plant diversity; moreover, human/cultural-ecological interactions further expand the topic of agriculture to relations that are commonly less considered.

The "Real Dirt" of Agriculture

One need not even look as far as rainforests to find evidence of a problematic effect modern industrialized (but also some traditional) kinds of agriculture are having on the very foundation of agriculture, soil. As undervalued a resource as soil is, treated as if it were a mere matrix for supporting plant growth, the fertility of which can simply be replaced with artificial fertilizers, it actually is more than a simple mixture of minerals and organic matter: it is a veritable ecosystem in and of itself, providing various essential ecosystem services.

Not only is the replacement or even enhancement of soil fertility by mineral fertilizers eventually unsustainable, at the very least as their supply from or with fossil resources (as raw material and/or source of energy) is finite and actually a cost factor; it is even less understandable considering that ecological processes are sustainable and essentially for free. Even now, natural processes contribute 90 percent of the nitrogen that is taken up by plants (Daily, Matson & Vitousek 1997: 124f.), but we are losing soil itself to erosion and experiencing its degradation caused by salinization, nutrient depletion, compaction, and other processes. The corresponding numbers are rather uncertain, but it is estimated that 17 percent of the vegetated land surface has undergone human-induced soil degradation since 1945 (Oldeman 1994), and it is to be suggested that conservation of soil, and measures to maintain and improve its natural fertility are sensible and indeed necessary considering both shorter term economic analysis and longer-term sustainability.

The inevitability of a focus on soil conservation becomes obvious when one takes the time frames involved into consideration. Erosion can wash large amounts of fertile topsoil away in a matter of hours and days, causing losses to agriculture as well as problems with sedimentation and siltation affecting hydroelectric dams, irrigation systems, as well as fishing areas off shore. Yet, in view of its buildup, topsoil can only be considered an ecological-geological inheritance: Physical weathering of rock and life processes work together over times of about 100–200 years to form any productive land (topsoil) from rock at all, and 200–1000 years to (re)generate just 2.5 cm of lost topsoil (Tilman 1988: 214–216, Pimentel et al. 1993, quoted in Daily, Matson and Vitousek 1997: 128).

For understanding the eco-economic concern with conservation of soil further, Daily, Matson and Vitousek's (1997: 119) comparison of the mere value of physical support for plants provides an excellent example: In agricultural balance sheets, this value which soil in agroecosystems could provide virtually forever if sustained is not counted, but the cost of replacing it with technological means, for example the "physical support trays and stands [in a hydroponic system] amounts to about \$55,000 per ha ... [and t]hese require replacement after perhaps a decade of use."

Thus, even just considering physical support in the time frames involved it is clear that we are living off a capital that is eventually irreplaceable or so costly to replace with limited technological means that conservation is essential.

Soil, however, provides additional services such as the nutrient (e.g. nitrogen) availability already mentioned above, i.e. soil fertility. In ecologically functional, biodiverse soils, this service can be provided essentially for free by fertilizing with leguminous plants (also usable for fodder production, or in intercropping), and by "biotic regulation" (Swift & Anderson 1993).

By this process, dependent on biodiversity of soil biota interacting with vegetation providing litter and root exudates, global biogeochemical cycles of elements are regulated in soil (cp. Daily et al. 1997), and nutrients held in "resistant" forms stored in the soil are made available to plants when needed through soil biota-vegetation interaction (Swift 1997, Swift & Anderson 1993, quoted in Jordan 2002) rather than washed away unused to as considerable extent as inorganic fertilizers are.

The work of soil biota has further regulatory effects on the hydrological cycle through its influence on soil tilth (texture and related properties). Tilth and organic matter content largely determine the availability and storage of both oxygen and, especially, of water (Jordan 2002). Together with vegetation cover and its root system soil structure further influences resistance to soil erosion and promotion of water infiltration, and even further-reaching aspects of the local and supraregional hydrological cycle (Daily et al. 1997: 117f.)

In a wider perspective, soil services can be seen as including the landscapes it creates, the ceramics that can be made from clays, and even the cultural diversity associated with soil, agricultural techniques, and cultivars (Daily et al. 1997: 126).

In most circumstances, "alternative" forms of agriculture provide preferable dynamics in regard to soil conservation and fertility than industrial-conventional ones. Soil fertility, structure and tilth (which decrease soil erosion and increase moisture uptake and retention), and microbial structure (diversity and activity, which promotes fertility and tilth) have been shown to improve under organic agriculture (Rodale Institute 2000, Maeder et al. 2002), for example.

Agricultural sustainability therefore tends to be more likely under such regimes of cultivation. Local circumstances need to be strongly analyzed and utilized in planning, and for improving experience and adapting agrisystems accordingly, however. In this regard (which will be found time and again), sustainability requires the (re-) development of a true "knowledge society" in which global concerns and aims are translated into local practices and adjusted in a co-evolution of practice and experience working with the immediate environment.

One of the persistent patterns of such a process, which makes its difference to industrialized models stand in particularly stark relief, is the occurrence of a large diversity of approaches being tried out (cf. Durham 1991), the very fuzziness, i.e. diversity and flexibility of cultivars and methods (cf. Nazarea 1998), of which accounts for the resilience of traditional agricultural systems.

With agriculture (and agricultural science, like ecology) being an experimental process rather than an exact science, organic agriculture, too, does not necessarily fulfill all suggested factors of sustainability per se (neither does conventional agriculture never meet any of those requirements, cf. Pacini et al. 2003). Even just considering soil erosion, for example, interactions between environmental conditions (such as climate, soil structure and topography of fields), agricultural technology (e.g. intercropping, existence of vegetation cover, terracing), and plant species and varieties utilized (e.g. annual/perennial growth, development of root system), need to be taken into account.

Taking sustainability further, into the relations between the agroecosystem and patterns and processes beyond it, "wild" and human ecological alike, further indicators and practices need to be considered even more strongly. Nevertheless, both the adaptive potential of a diversity of approaches, and their orientation towards sustainability, is promising and needed to be promoted further.

Wild Relations

In addition to the base of agriculture in soil, there exist further ecological relations which require the orientation on integration, both of agricultural activity with ecological processes at global and local (and even species ecology) levels, and of agricultural lands with areas that are still more natural or wild. The coexistence of agroecosystems with wild habitats and structures such as hedges and fallows promoting dynamics amenable for sustainability is especially relevant as an example for those concerns, e.g. of conservation mentioned above, to which organic agriculture does not necessarily contribute in and of itself; its benefits have to be understood, and it has to be made an objective. The best examples for the utilitarian value of these dynamics are provided by the ecosystem services of pollinators and of natural pest control. Less immediate relations concerning the import of wild region's resources to nutrition and of wild biodiversity to potential innovations in/for agriculture will also be addressed.

The services provided by pollinators, including wild ones (and wild faunal diversity in general) are among the least valued, while their provision is intrinsically linked to the need for conservation of wild habitats fulfilling these species' requirements of places for nesting, overwintering, foraging, etc. (Nabhan & Buchmann 1997).

The very fruit set of plants other than wind-pollinated crops and (to some extent) self-pollinating vegetables, most obviously of fruit, for example, depends on the service of pollinators. Some of the insects providing this service are domesticated, i.e. honey bees (also) kept for honey and wax production, but wild ones, too, are necessary and contribute to the resilience of the system of pollinators and pollination-needing plants (which of course includes non-cultivated ones as well). Among other things, conservation efforts for threatened plants may fail if pollinators are not protected; so the effect that the decline of (wild) pollinators has amounts to both anthropocentric-economic (agricultural) and ecocentric-ecological (conservation) trouble, e.g. a "pollination crisis," with an associated reduction of agricultural productivity (cp. Nabhan & Buchmann 1997). Even disturbances of historical and ongoing evolutionary mechanisms (Buchmann & Nabhan 1997, Wilson 1992) and changes in ecosystem structures and flows (Howell 1974; Heithaus 1974; Roubik 1993) are associated effects.

The way insects are more immediately thought of would be as pests to agriculture, but far from being totally controlled through the farmer's care, a large proportion in most of the world is even now not controlled by pesticides but by natural processes, both the service provided by natural enemies of pests, and "climatic-related controls ... that interrupt herbivorous pest reproduction cycles" (Naylor & Ehrlich 1997: 151). The former of these are, like the occurrence of wild pollinators, again related to the existence of sufficiently diverse, wild habitats providing shelter to pest predators, but also to the existence of the pests. In contrast to the use of pesticides, these controls may appear less efficient in a narrow, short-term sense – they are oriented not on the extirpation of pests, which would ultimately be impossible, but on a dynamic balance shifted in favor of crop production. Thus, however, they impact less on other ecosystems, helpful species, and human health, while not being susceptible to pest species becoming adapted to the pesticides (which is regularly the case, if not the norm). Moreover, such controls are essentially free or, except for potential initial cost for their establishment, more cost-effective than the purchase of ever new, more costly pesticides (cp. Jordan 2002: 158f.); plants with built-in protection, i.e. genetically modified to produce toxins against pests may work, but probably will result in an adaptive pressure easily realized by the insects affected, particularly as the rules for keeping patches of unmodified plants (in order to make for less evolutionary pressure and a higher chance of continuing interbreeding between adapted and non-adapted pest populations) are not followed.

Ways in which "natural pest control services" (Naylor & Ehrlich 1997) could be promoted include the use of intercropping, companion planting, and other forms of increasing the diversity of crop species and varieties in order to enhance the resistance and resilience of the agroecosystem. This can be further promoted by working in conjunction with "wilder" areas: A push-pull-system of combining such intra-field diversity and deterrence ("push") with areas planted in species more attractive to the pests than the actual crop ("pull") has been suggested and tried out (cp. Parrot & Marsden 2002).

Taking the view on the relation between agricultural and wild ecosystems further, other concerns surrounding the effects of intensive industrial-type agriculture on both natural ecosystem and human health, and the reverse impact of the deterioration of wild ecosystems and their services on agriculture, comes into play.

Agrochemical use, for example, impacts on both kinds of health, ecosystem/environmental and human, both directly and indirectly. To expand on this case already mentioned above, the input of nitrogen – the anthropogenic input of which now equals or surpasses the size of its nonanthropogenic cycle – has such an extent for the perceived advantages it provides, but a less general view is in order: A large extent of the input is washed or leached away, representing an unnecessary expenditure and a factor capable of adversely affecting both ecosystem health, e.g. the biological control in soil (as noted above), and human health. Its effect on further ecosystems, e.g. rivers and coastal areas, is that of eutrophication which translates into environmental deterioration as well as changes in the human nutritional situation as fisheries are affected.

In a related kind of dynamic, extending to other forms of e.g. urban, industrial, etc., runoff, pollutants and novel substances also "come back to haunt us" by accumulating in the food chain; fisheries, because of the connection between their importance for human nutrition and the particularities of toxin distribution in water and their (re-)accumulation in their course through the oceanic food web are a particularly good example here, too.

The further impact of excess nitrogen on natural ecosystems (cp. Moffat 1998, Kaiser 2001, Nosengo 2003), eventually falls back on agriculture if changed functioning or deterioration of the wild ecosystems translates into changes in regional and ultimately larger-scale cycles such as the hydrological one, changing the amount and variability of water availability. Putting it the other way round, and into a less negative perspective, it can be argued that agriculture, although it always has to imply a change in land cover, has typically always been integrated in the wider ecological matrix of its environs, and profiting from – or even just functioning by working with – the patterns and services provided by this integratedness. This applies similarly to the large-scale of hydrological relations between forests, watersheds, and agroecosystems (which can even work like seasonally aquatic systems, particularly in the case of rice), as well as to the smaller-scale, as of swidden fields within the forest, and/or plantations of valuable crops within forest areas with largely closed canopy.

Apparently, such considerations imply that eco-agricultural approaches could not only solve the one basic problem they were aimed at dealing with, but could exert an even stronger positive effect. Further synergies arise when/if an even more comprehensive eco-cultural approach can be employed and implemented in practice. For example, as noted above, there would be synergies between the cultural aspect of preferred diets, especially if support for a healthier diet could be raised, the lower intensification necessary to provide for such a diet, and the potential of eco-agriculture to provide the foodstuffs of such diets in a way that is more likely to be sustainable and adaptable where changes are necessary.

Another role of wild habitats – and therefore of their conservation – lies in their relation to the biodiversity of crop relatives, potential crops, and wild species utilized in human diet, all of which have an effect on the sustainability – both in terms of food security and of human well-being – of the food system. The further interrelation with biodiversity will be considered in a following section, but now is a good time to turn to the meaning of wild foods.

As a matter of course, foods harvested in the wild play less of a role in a rapidly modernizing, urbanizing world where many children may grow up believing that their food came from the supermarket. Yet, a closer look reveals their continuing importance.

First of all, from a more traditional anthropological perspective, contributions to a population's nutrition that come from hunting and gathering do in fact play a role. Etkin (1994: 4f.), for example, notes that "wherever the consumption of wild foods has been accurately assessed [both among the traditional 'objects' of anthropological analysis and among agricultural populations], they emerge as regular and important elements of diet." In this same context, the relevance of wild plants as "emergency foods," which would not normally (or only as an aside) be grown, but utilized when primary crops fail, could be revived or taken as a lesson for the importance of diversity and the potential use to which some "weeds" could be put.

In the case of "developing" countries, this perspective draws attention to one of the potential roles for wild habitats as repositories of valuable or emergency products in times of need – as they are commonly utilized even today, unintentionally as when the needs or very existence of local populations is not considered during the establishment of national parks, and in more promisingly directed, co-developing ways in the more recent approaches to biosphere reserves which may be used in sustainable ways. The problems of over-exploitation of commons, particularly in situations of rising population and disruption of traditional social patterns of control over them do strongly apply here, of course.

In an emergent modern approach, e.g. of sustainable agriculture, this wild diversity within the agroecosystem could, for example, be of use if ways are found to make it fit in with the ecological dynamics of fertility and pest control noted above, in the revitalization of landscapes that probably would make them more attractive to both human, tourist or permanent, and other ("wild") inhabitants, with a potential role in rural development and for a balance between agriculture and (sub)urban sprawl. In most cases, the promotion of developments that move in such directions would depend on recognizing and supporting the multiple benefits beyond food output obtainable from farming. For the more entrepreneurial of farmers (especially if changing support structures would take some of the initial cost and risk away), such diversity could also be of profit if utilized in on-farm production of specific products.

Similarly, a removal of the distinction between wild and farmed, urban and rural could turn out as promoting conservation of natural/wild habitats in the end if it makes land conversion less of a necessity, the fulfillment of basic and "biophilic" needs better understandable and promote-able as requiring both the human-dominated/farming "wild" and the non-human "wilderness" (and, it would make the above-mentioned conservation of wild habitats as both a conservation measure and a repository for emergency foods, etc. usable). This development would be supported by urban and community-supported agriculture, for example. Urban agriculture, incidentally, appears to be relatively common (still, or again?) in "developing" countries as rural farmers migrate to cities but find no other way to make a living than that (Bryld 2002). In cities of "developed" countries, on the other hand, urban and community-supported agriculture (CSA) are struggling but promising developments; they contribute to nutrition, personal agency (outdoors, hands-on activity), values and knowledge/education of a green and socially-oriented kind (the latter in relations between farmer and clients in CSA, or among residents of a neighborhood getting together in community gardens, for example).

The importance of truly wild habitats in and with agroecosystems in the modern context has another component to it that should not be underestimated: the special position products from the wild tend to hold with many people:

Hunting and fishing, however disliked by some, are one such way that provides food and/or fun and incomes, as do non-consumptive activities like birdwatching; wild plant, fruit, and mushroom gathering. This addition – garnish, aromatic, healthful contribution – to diet and/or pleasurable outdoors activity has a special appeal to many, particularly the more removed from such experiences ordinary life has become, too.

Finally, the special position that their status of being wild-harvested imparts to certain products (which is particularly strong with regard to some herbal medicines such as ginseng, for example) that makes them exceptionally prized should be noted, although it does not go without adding that market pressures tend to promote unsustainable profiteering in many such cases.

All of those meanings ought to be brought to the consideration of sustainability; some of them could be applied directly, some as a factor to promote different politics and economic orientations.

Food, Agriculture, and Biodiversity

The diversity of life is a rather well-recognized concern of conservation biology, argued to be in the middle of another historic downturn (this time not caused by a cataclysmic meteorite impact but by compound side effects of human numbers and ways of making a living). The deep dependence of human on other life is most apparent in the case of nutrition, and although it intrinsically contains an element of exploitation – after all, the plant composition on a piece of land usually has to be changed to reap a sufficient harvest of the preferred edible plants from it, animals must be killed to gain meat – it provides yet another perspective on the positive interaction that can be attained. The diversity of species is a natural source of human food, after all, and the diversity of current cultivars of plants and races of domesticated animals has arisen with the influence of human choice working with that naturally occurring "material" – which is the artificial selection that actually gave Darwin his idea of the similar but non-conscious natural selection.

Several thousand species of plants would be edible, but only about 150 are or were used in world commerce, 30 feed most people, and only "the four major carbohydrate crop species – wheat, corn, rice, and potatoes – feed more people than the next 26 most important crops combined (Plotkin 1988: 107). The concentration is further continuing with the focus on fewer (under ideal conditions) high-yielding varieties, and even less genetic diversity among GMOs; the same applies to the diversity of domesticated animals, and even to industrial forests planted with clones of single tree species. Considering the evolutionary dynamics of diseases and pests, or the relation between ecosystem (including agroecosystem) resilience and diversity, this approach is hardly sustainable.

On the other hand, an extraordinary diversity of cultivars and domesticates has (been) developed in their "anthropo-biological" evolution (by artificial, i.e. human, selection). As may already be clear from some above comments, e.g. that on rice-growing on the Vietnamese-Chinese borderlands, this diversity not only exists as an absolute number, but is to be found even within smaller areas, and has been utilized within single fields.

For one, the cultivation/domestication of species/varieties within certain regions, with some but not much exchange over long spans of time has naturally produced (land)races suited to the usual range of conditions encountered in the particular region of cultivation. Additionally, the varying needs and conditions different varieties of plants need for their growth form an insurance against the vagaries of weather and make best use of different soil conditions. Furthermore, diversity is not just an insurance against negative effects, preventing the total loss of harvest in all but the most dramatic of conditions (when the above-mentioned emergency foods would have come into play), it can provide further benefits such as disturbance of pest cycles, restriction of weeds and diseases (through allelopathic effects), overyielding (Jackson L.E. 1997, Wolfe 2000, Zhu et al. 2000), and positive effects on plant protection and soil fertility (Altieri 1999).

Secondly, the diversity of properties provided by such a wide range of foodstuffs translates into a diverse diet likely to contain all the necessary nutrients, including micronutrients or secondary phytochemicals that are of further dietary and even of outright medicinal effect (see Chapter 15). The traditional Amerindian combination of corn, beans and squash, for example, forms a basis for a balanced diet and works to fertilize itself (by way of the legume's nitrogen-fixing ability), control weeds and erosion (through the ground cover provided by the squash), and counteract diseases and pests (Peña 1999: 121; cf. Altieri et al. 1987).

Apart from the relation between biodiversity and agroecosystem functioning – both processes (of pest cycles, soil fertility) and patterns (of more or less diverse, cultivated or relatively wild habitat-fields) – which has been described already, a further relationship exists between crop and wild food plant diversity, concerning the potential for agricultural progress. Locally adapted races and wild relatives of widely cultivated plants are extremely valuable, for example, for local importance, as new crops, or to improve cultivars:

First of all, they exhibit a potential for satisfying local food needs in relatively self-sufficient or idiosyncratic ways. The former applies to fringe situations where high-yield varieties cannot grow well because the conditions they would require cannot be provided, for example. An orientation on such a condition should rather make it the norm for sustainable conditions taken seriously, as resilience – sustainable yield – is more necessarily a factor than mere yield. The latter aspect of landraces' "idiosyncratic" value is realized where culturally-mediated preferences for foods that may not be staples of the global food economy are effective (and, of course, where these two factors come together).

Secondly, landraces could turn out to be "new crops" which only had local relevance so far, but could be of interest as a tradable agricultural produce or as region-specific product (which interest is best known in wine, which can hardly be considered new, but can be found with many old– chile peppers, chocolates, i.e. cacao varieties, tomatoes – and potentially with new cultivars, e.g. quinoa and amaranth grains).

Finally, wild relatives and landraces of cultivars are valuable for improving properties of agricultural varieties, whether by traditional methods of plant breeding or biotechnological ones (cp. Iltis 1988, for example). Unfortunately, the economic and discursive clout of the latter, with its potential for patents and royalties, belittles the value of landraces. Conservation of traditionally used/developed cultivars is not only not being considered as progressive in contrast to biotechnology, but as an ongoing evolutionary process requires, as Soleri & Smith (1999) note, that the farmer be working as breeder.

The consideration of the "anthropo-biological" diversity of varieties and races not only relates to current and future global food security and sustainability, but also, with the interaction between food, food systems, and people, to factors such as cultural identity, health, and human well-being, which are more immediately relevant to the individual person.

Agri-Culture

Food is also the most obvious and open-to-experience factor connecting culture(s) and social groups, the individual person, and the environment in their diverse interrelations, from general processes at large scales to the individual's embeddedness in local environments of both nature and culture. To many people nowadays, the conditions of food production are rather far away from the factors immediately affecting their lives – although farming remains a fundamental way of subsistence and life for many others – so that the concerns described above are not necessarily high up on the agenda. On the other hand, environmental and agricultural sustainability and the problems of agriculture, whether food security, contamination from pesticides, or farther-ranging ecological impacts, do affect the conditions under which everyone lives.

Focusing not on proxies like GDP growth but on more direct measures of food security and sustainability and the fulfillment of people's need for an adequate diet, the divergent concerns of too little and too much food are more connected than they appear. – In both cases, agriculture, ecology and culture in their less considered relations and in their context of politics and economics (e.g. support for different developmental pathways) are the major factors, and agroecological and eco-cultural approaches exhibit promising dynamics: The promotion of eco-agriculture guides agriculture towards food security which is primarily based on the ecoregion and only secondarily on trade relations and food buffer stock, and towards sustainability by promoting the interplay of ecology and farmer's knowledge and creativity rather than dependence on capital-intensive input of energy and (ultimately limited) materials. From this basis, the status and potential of rural land and its inhabitants could be improved. After all, a major part of the world's population is or would still be working as farmers providing food and work for themselves and their communities, so their empowerment is one step to better the situation of the poor. Moreover, farming in industrialized countries, in a form that is empowering e.g. by reducing input-dependence, is a possibility to improve farmer's situations and make it likely that their contribution is more rightly valued.

Through an integration of farming, rural communities, and "wild" habitats, e.g. with CSA or sustainable agriculture connected to local enterprises making products from it (or doing so on-farm), the situation could be further progressing towards better living conditions and sustainable ways. – Examples have been collected in the USDA-SARE's "The New American Farmer" (Berton 2001), as well as being described in Lappé and Lappé (2002), Hawken, Lovins & Lovins (1999). Another interesting example worth of being pointed out is "Mary Jane's Farm" (see www.maryjanesfarm.com) where organic produce is being turned into dehydrated menus, having started out as an enterprise for backpacking food and since expanded into organic, vegetarian fast/convenience food (which oftentimes appears a contradiction in terms), and more, supporting a number of people and a chosen lifestyle. Erich Stekovics in eastern Austria (see www.stekovics.at) is another example for how organic (?) agriculture of old cultivars – in this case over 1200 varieties of tomatoes – can be a profit to farmer and consumers. (The supposed diversity in the supermarkets, particularly in the agricultural produce section, on the other hand, is a paragon example for what Barber, 1995: 116, describes as "McWorld's strategy for creating global markets [much of which] depends on a systematic rejection of any genuine consumer autonomy or any costly program variety – deftly coupled, however, with the appearance of infinite variety." A pleasurable diversity of species, varieties and tastes falls by the wayside.)

The higher-scale concerns of ecology and sustainability and their meaning to the farmer and society have been addressed in the course of this chapter; "culture," as ways of life that include what foods are prepared in what ways, eaten in what situations, and so on, has been mentioned there as a major factor affecting the nutritional (and health) situation, already. Even in situations where they are not usually considered, however, the essentiality of eating, its pleasures and problems, relations to individual health, and role in society and culture, do make it an excellent vantage point for considering the potential relevance of a positive ecology of agriculture/nutrition to each and every person, not just in long-term perspective (of survival) but in life practice.

First of all, nutrition, in the case of obesity in overdoing it, is probably the best contemporary example for the importance of an educative-empowering individual-/person-oriented perspective, requiring that the biological and cultural predispositions, knowledge, and conventional ways of life, and individual agency and action, be balanced; contexts such as the "toxic food environment" (Brownell & Horgen 2004) are exerting tremendous influence on all but the most strongly inclined to rigorous individualism of food preferences (or pressured to it for health reasons), nonetheless, so that political and economic changes will, once again, be essential as well. In this case, on the other hand, positive examples have been initiating change, and the advantages to the individual, communities, and society at large should support it further.

Enough food is a necessity of course, but considering the adequate composition of a healthy diet or the pleasure that lies in diversity vis-à-vis the supposed requirement of giving up pleasures to "save the planet" is a necessary lesson: Eco-agriculture would be well suited for providing a diet oriented towards crops, vegetables, greens, and fruit, only supplemented by high-quality meat, rich in diversity of foodstuffs, and this in turn is closer to what is considered a healthy, balanced nutrition, as well as supportive of attempts to "feed the world" (as eating lower on the food chain requires less input).

Moreover, this approach to food provides a new way of taking control of one's life at least in this area, choosing a health-promoting and pleasurable way of eating and of recovering the feeling of being in the world. For one, regionally grown foods, and community-supported agriculture even more so, connect the person to the total – natural and social – environment. A more active engagement, for example by CSA, "self-harvesting," or community gardens, can further strengthen this link, as well as provide an opportunity for activities beloved by many people. – The positive effects of green surroundings and practical engagement with nature (both wild and human-created) will be encountered again in the course of this work. Among local and indigenous groups, community and/or traditional activities connected with agriculture, hunting, and gathering still function as a major social, identity-building activity. Providing food turns into a secondary (if any) concern of such pastimes, of course. Yet, urban agriculture has been argued to be relatively widespread as a way of gaining subsistence in developing countries (Bryld 2003). Still, their potential for being both practical, emotionally worthwhile, and educative is great.

Health is not necessarily the best example, yet, as this regard is oftentimes co-opted for marketing "silver bullet"-solutions such as functional foods and dietary supplements, and rigidified into ideologies. Some traditional forms of nutrition, particularly when diverse and (therefore) naturally rich in phytochemicals are less supported, but balanced with lifestyles (e.g. physical activity) more likely health-promoting; but that is not the only factor.

After all, food is a basic need as well as a major field of personal, cultural, societal, and environmental relations: What you eat is, as the saying goes, what you are. Certainly, how foods are typically prepared and in what combinations does in fact make for their particular regional or ethnic character (cp. Rozin 1992 [1983], for example). Where and how people get together (or do not) to eat means a lot for personal and community identities, also, as eating commonly has a social dimension.

Knowledge, whether gained by formal education or embedded in social learning, which has been mentioned as a decisive factor of an agriculture that is oriented on sustainability, is just as pivotal in these relationships. Far from being a dry matter, however, knowledge of·the diversity of foods, even among single food species in the form of their "biocultural" diversity of cultivars, finds its expression in sensory experience, in the wealth of tastes. The Slow Food movement aiming for "the defense of the right to pleasure" (or good taste; in the original Italian: "per la difesa del diritto al piacere") is a better promotion of sustainability than most environmentalist/conservation organizations where the dominant discourse may make it appear as if the nature/environment they try to support had little to nothing to do with human lifeworlds, especially in a good, pleasurable way. Yet, as Clif Bar Inc. (online) simply puts why it has turned to mainly organic ingredients for its performance bars (another example of an interesting enterprise supporting organic agriculture): "Because it's better for you. ... Because it's better for the planet."

Chapter 12

Clothing:
(Not Just) Temperature Regulation

As the need for nutrition is mainly satisfied by way of the creation of agroecosystems, the environments (considered) capable of satisfying it the best, so the even more immediately lived-in environment of human beings is typically an influenced and even created one. Where other animal species have developed physiological adaptations to deal with the need for regulation of body temperature, and do not even feel a need to create constructions that provide protection, to feel physically secure, in many cases (although a number of them build constructions that could inspire, and humble, great architects), human beings invariably feel these needs and satisfy them by clothing themselves adequately and building some kind of homes. The problematic nature of this obvious human universal is just that obviousness: Both architecture and, in particular, clothing/fashion (where the impacts are even more spread-out and not as visible immediately) are typical human concerns situated in the social-cultural sphere to such extent that their relationship to the environment by and large goes unnoticed.

Both elements of how the needs for temperature regulation and physical safety are satisfied, like most needs, immediately go beyond the mere basics of their fulfillment. Rather, they are influenced by (natural and economics-based) availability of materials, traditional and individual preferences, and so on. At the same time, their "ecologies" of natural and cultural environments, individual and group agents, hold lessons for how human life may be better integrated into ecospheric functioning. Interestingly, however, the intrinsic relation between (these) human needs and natural environments has been further and further moved apart in historic time, mainly as the raw materials used are transformed to a large extent by (high) technology rather than extracted and used locally and with little manipulation. Yet, neither are the effects of our construction of conditions on the natural environment difficult to understand (not least with the impact of energy and materials used in their production and upkeep, and the problem of their disposal once considered wastes), nor are their relations with ecospheric functioning and environmental conditions nugatory, at least as climatic conditions determine the background against which they are operating.

The need for temperature regulation may be the ultimate rationale for clothing, but many further relationships impact on it, so that this issue can be used for the more general discussion of the modern economy's relationships with the environment on the one hand, and with human needs and culture on the other hand. After all, more than just to protect from the elements, the proximate reason for which people dress in certain ways is the requirement to be dressed adequately, to express social and cultural belonging or personal identity, to follow fashion or make a fashion statement oneself. Typically, neither the way in which clothing was produced, both in ecological and in social respects, nor their ultimate suitedness for fitting into ecological cycles is of too much concern.

The same applies to most, if not all, other amenities and technological toys of modern life which are wanted because they are seen to satisfy some need, suit some primal diversion, make life or work easier, etc.: Environmentalist considerations are very low on the list of factors to take into consideration for all but a few consumers. From a psychological or religious, as well as an environmentalist/ecological perspective, it may be good to come around to less unconditionally enthusiastic, more reasonable attitudes to consumption, e.g. where it is used as replacement for social needs or arguing with advantages empirical analysis do not show it as having (cp. Kasser & Kanner 2004). But, of course, emotion is a strong, natural characteristic of the human makeup so that such a "need" (that is, a liking) for consumption will not go away.

With clothing at least, whether because it were a necessity or because its environmental relations go unacknowledged, calls for doing without to save the world (comparatively common as they are with regards to technology for entertainment) have not usually been raised. This makes it easier to focus on the requirements of integration to move towards sustainability, at least, and to consider the need for a comprehensive perspective, working with those who would see their advantage in sustainable ways of life, those who would produce a change in techno-economic systems, and towards wider "cultural" change. The issues to consider in this context, consequently, are those of materials sourcing, production, and cycling/waste (i.e., technology and economics) on the one hand, and those of needs and culture, on the other.

"Integration" in the case of technology and economics refers to its re-orientation to work like ecological processes, as suggested in ideas of "industrial ecology" (cp. Benyus 1998) or the concept of "natural capitalism" (Hawken, Lovins & Lovins 1999). Their major thrust can be described as shifting from a linear model of economics bringing resources into the human sphere, transforming them to create value and make a profit – in many cases, even by creating the need for a product rather than a product to fulfill a need (which accounts for why "the economy" is supposed to always need to grow while people may not) – and finally dumping them back into "nature," to a cyclical model that must not know externalities or wastes, but work with substances and in ways that will fit in with and mirror "natural" (non-human) ecological functioning (cp. McDonough & Braungart 2002).

Challenges that arise include, for example, that even the renewable and reusable materials that would eventually only fit such workings needed to be produced in similarly "integrated" ways, e.g. eco-agriculture. Non-renewables could suit the sustainability criterion of intergenerational equity if the gains achieved by their consumption are channeled into the development of new, alternative sources by the same amount, cp. Goodland 2003; however, they do not fit in with considerations of impact in terms of ecospheric functioning/cycles well, if at all.

Moreover, no novel substances should be introduced (at least into the outside of the industrial site of production) that could exert detrimental/toxic effects, (bio-) accumulate, or not be safely or economically returned into the ecological or productive-industrial system, respectively. This does imply a need for radical, but far from impossible, changes if one is to judge by the suggestions and even cases of innovative approaches one finds in the literature (Benyus 1998; Hawken, Lovins & Lovins 1999), and certainly strengthens the case for eco-entrepreneurship. The economic verdict presented by those cases is not yet entirely clear, but seems to offer hope that such approaches could even be economically competitive, particularly as long as a premium for more "eco-friendly" products can be achieved, when externalities are not factored in; not allowing for externalities, as well as keeping a longer-term perspective – on both human and economic prospects, for that matter – in mind, the verdict that only a non-disruptive, ecologically restorative, sustainable way of life and way of making a living, including economy and technology, can be the future should hardly be objectionable: everything is dependent on a functioning ecosphere, after all.

The prospects for economic growth under conditions where labor is ample but natural capital increasingly scarce (cp. Hawken, Lovins & Lovins 1999) and where it may be necessary to make it parallel a climax ecosystem, are rather more difficult to ascertain.

With ways of accounting that do not count environmental destruction as income and the associated loss of ecosystem services not at all (as they are seen as externalities), the restorative economy reinvesting in natural capital (one of the criteria of "natural capitalism") would be counted as growing, new products or probably preferably services, including a potential economy of knowledge and action such as around wellness, learning, "personal growth" – in short, the already growing economy for "LOHAS – lifestyles of health and sustainability" (cp. Ray and Anderson 2000), may provide further opportunities.

Eventually, however, the ideal of infinite growth must probably be given up (cp. Daly 1997). Questioning the ideology of growth, as proxy for satisfaction of the variety of human needs, would be a worthwhile endeavor (not just academically or in activism, but more widely), particularly as a failure in terms of the current measure of economic growth would definitely not mean the end of progress (nor the end to the need for progress), e.g. in satisfying needs and improving well-being. Nor, to counter the usual ideological argument, would or should trade and profit (and many, if not most, people's urge for it) disappear, although it would – need to – be moderated, at least with a consideration of sustainability. Nothing less than such a directing is what politics or other social/cultural mechanisms guiding behavior into accepted paths are there for. Unfortunately, they are currently rather ineffective both by working without as strong a reflection on the need for change towards sustainability as necessary, and by giving overly strict specifications on current (almost) best practices that smother inventiveness rather than change the system of support and penalty to promote eco-entrepreneurship.

Fashion, and the consumption going hand in hand with the wish to stay en vogue – as an example standing for other factors influencing (making?) the consumer society – would not be quite such a problem if technology were mimicking and integrated into ecological functioning in such a way as to be "integratively" sourced and allowing (and for economic reasons requiring) for "wastes" to be safely returned to economic or ecological cycles. As was the case with the problem of overconsumption in nutrition, so individual lifestyles and their aggregate of culture nevertheless remain an issue. – The exclusive focus on money on the bank and consumption of goods as the sole measures of a good life, at least, need to be challenged. A standard of living can be measured in such a way, but quality of life and well-being or happiness are related to consumption/economic growth to a lesser extent (cp. Eckersley 2000). This understanding can contribute greatly to the perspective of positive ecology, i.e. to how an orientation on sustainability can actually enhance the quality of life, even in cases where it entails a reduction in consumption.

For example, buying high-quality, durable clothing, preferably produced in ways that are ecologically better (e.g. organic cotton) and socially just (e.g. from fair trade), is an expression of personal identity, on the one hand – a reverse fashion statement, so to speak, if the usual sense of that phrase represents the ongoing hunt for what is currently in fashion. Additionally, it need be none the more expensive than cheaper, lower-quality products that need to be replaced more often, but could be better for the person (in terms of intrinsic self-image and comfort/health if well-working materials and products are made without substances impacting on health) as well as "for the planet." Such suggestions, interestingly, can already be found in a context not only concerned with sustainability but related to life simplification and similar ideas for creating a more individually-shaped good life (e.g. "voluntary simplicity," cp. Elgin 1993). On the other hand, it is worth noting again that this is not a straightjacket only way of life that everybody would have to adopt (particularly not in the same way), but only one suggestion to consider; if the economy worked in more "integrated" ways changing what is currently waste into a reusable/recyclable material sensibly treated in such a way – which would be a preferable approach to the durable products as well, after all – higher levels of consumption could possibly be sustainable, as long as they are more equitably distributed and not exceeding the scale of ecological limits.

The potential of a balance of having enough but not striving for ever more, and a focus on more than just possession of goods, is a particularly necessary consideration to counter the supposed sacrificial altruism (giving up of amenities) sustainability/environmentalism is commonly seen to entail. This holds true in particular in those cases where material consumption represents an attempt to replace the lacking fulfillment of needs which are actually unrelated to material standard of living, e.g. sociality, a sense of agency or meaning in life, or – most strongly in fashion and fashionable, label, goods – of self-assurance. At the same time, the importance of such emotional issues as social justice (which sustainability is, after all, meant to entail as well) on consumptive choices should not be denigrated; wider needs/concerns do tend to be a major motivation for particular ways of life, probably more so than an utter rationalization of sustainable consumption as lifestyle (cp. Hobson 2002).

Chapter 13

Architecture:
Temperature Regulation and Safety

Pointing out the network of relationships surrounding how humans and their homes and habitats are related seems somewhat preposterous – it would seem to be readily apparent, too. Yet, especially in the context of sustainability, only the way individual buildings are constructed and styled is usually considered, even though much wider contexts are of relevance: As a matter of course, although temperature regulation and the provision of physical safety may have been the earliest raisons d'être for shelters and homes, socio-cultural factors, personal preferences, and technological capabilities, environmental conditions and landscape patterns, and the rural or urban contexts all interact in this issue.

Under "traditional" conditions, sustainability – or the opposite of human impact on immediate environment and global ecosphere – has been less of an issue, including as concerns the built environment. Impact was mediated by smaller population and settlement sizes and availability of regionally sourced, natural materials only in the majority of cases, the odd exceptions of monumental constructions and long-inhabited cities notwithstanding. The changes effected by the latter (although far from absent, even if only by being more localized) were not quite as detrimental as those now arising from still-increasing city sizes and sprawl, consumption and the drive for its increase for the sake of development, elevated by technological prowess and happening on a global scale, have been coming to be. In poorer, more "traditional" conditions, the situation still is similar, i.e. many buildings are still made from little modified, regionally sourced materials – although again, their production does entail environmental impacts, and in some cases certainly too much, and new technologies offer new chances in addition to the novel problems they present. Yet again, the challenge is twofold: tempering the impact of modern, increasingly urbanized and technologized, ways of life, while making possible an improvement of living conditions – in part by and with technological amenities, and therefore by the same approaches as for the former, probably – where development in this sense is of necessity.

Particularly with the perspective set forth here, cultural approaches such as a rebuttal of the singular future seen in total technologization at least since the 1950s (a more sensible consideration of which could decrease consumption and impact caused by it) should form as much a part of the program as should acceptance of the likelihood (if not fact) that the possibility for obtaining better houses and a variety of technological amenities will be an option preferred by most, certainly those still lacking what are considered basics in industrialized countries.

In this regard, the orientation on an integration of economy/technology to function in and with ecological cycles by mimicking them, e.g. in processes of production (as described above) forms one major part of the equation: If the products making up those amenities could and would cycle within the economic system or be safely returned to natural ecological cycles, they would pose much less of a problem. The same comments made for clothing and technological amenities apply to materials and products in the context of the built environment similarly, albeit with (radically) different time frames and cycles being involved (ancient ruins, if not rediscovered and protected from nature for cultural reasons, are quite naturally a part of the Earth's geology more than anything else, for example). Secondly, however, the positive influence on and ongoing provision of adequate living conditions, e.g. – to return to those basics – a regulated temperature, but also the energy to run technological amenities, is a major consideration, and one that is contingent on constructions, individually and in their aggregate in settlement patterns, and on their relation to the (wider) environment.

Again, an orientation on sustainability (apart from not being supported and understood well enough) suffers most from its competition with dominant approaches subsidized through externalities, and currently higher costs up-front (unless, interestingly, in off-grid marginal situations). Contrary to usual presentations as requiring "sacrificial altruism," as having to imply a sinking quality of life, however, the suggestions brought forward for – ecological, sociocultural (including economic) – sustainability as regards the built environment are invariably oriented towards higher quality of life.

"Building Sustainability"

In the context of the immediate environment of homes and offices, in which a majority of time is spent, the admonition that preventing problems is better (in terms of health and of economics) than having to try to clean up later holds as true as it does with all environments, even if the conditions would be rather different: the built environment is created, designed, so it should be better possible to influence its character. Still, some of the combinations of architecture and materials that were considered the inevitable progress managed to create structures that produced the "sick building syndrome" in their inhabitants.

Examples of the design elements used to improve sustainability, or at least try to make architecture less un-sustainable, and that tend to improve the quality of the buildings as well as their construction and (especially) operating costs, include:

- Different approaches to working with the construction's physical orientation and layout in relation to the sun to promote natural lighting and "passive energy" standards that make heating less dependent on imports other than sunlight (even "plus-energy houses" have been built already); more energy-efficient and pleasant lighting systems are available, as well.
- Utilization of alternative energy such as solar heating and photovoltaic cells, which can even be integrated into facades, balconies, or as windows further contribute to making the building deliver its own (or more) energy.
- Cooling can be achieved through solar systems or architectural measures (as it had been in hot climates for quite a while) as well.
- Utilization of materials that are locally produced contribute to the economy here, require less transport energy while offering more possibilities (potentially, at least) for controlling the conditions of their production and looking for ones that suit all three above areas of design, but more importantly, choosing them with regard to freedom from harmful substances contributes to a healthy "building climate," and look and feel plays a further role in quality of life achieved.
- Relations with water, while usually little considered, contribute in various ways, too. Of course, the use of water-efficient appliances (or their replacement in the case of "ecological sanitation") plays one role; in "outside" environmental relations, possibilities for collecting rainwater to use in irrigation of green areas and provision of water management structures in landscaping, towards unsealed areas with plantings or even towards ponds as biological treatment plants (where possible), for example, can further support reduced impact and more pleasurable conditions, or even provide "natural" habitat.
- Larger buildings can utilize "waterscaping" even on the inside, with the potential of providing a better interior climate, as well as an interesting, relaxing (and unusual) view and soundscape.
- Plants can play a further role in ameliorating the look and feel of and around buildings, as well as measurable air quality and pleasurable atmosphere inside and micro-climate outside; in the case of "green roofs," the combined effect of a layer of soil and of vegetation is used as insulation against cold or heat (in the latter case, together with the cooling effect of evapotranspiration by the plants), and provides further "urban nature."

Design for sustainability requires seriously different approaches from ordinary architecture, among other things such that go beyond mere compliance with what law dictates and the designer or customer deems to be en vogue. Rather, according to the discussion by Williamson, Radford & Bennetts (2003) the different images that shape architecture (the natural, social/cultural, and technological) need to be brought into a "responsive cohesion" that considers the different orientations they entail together (and in that rather than the usual reverse or reduced order); achieving acknowledgment and mutual receptivity of different concerns by "reflective practice" based on "reasoned argument," which would show the skill of the designer in creating sustainable architecture.

It is well worth pointing out that, as far as the relation between architecture and its foundation in local conditions is concerned, the necessity of building from this basis has (or would have) been known since ancient times. Vitruvius, writing in the first century CE, for example, noted that, "if our designs for private houses are to be correct, we must at the outset take note of the countries and climates in which they are built. ... hence, as the position of the heaven [sun] with regard to a given tract on the earth leads naturally to different characteristics [of climate]... it is obvious that designs for houses ought similarly to conform to the nature of the country and to diversities of climate" (quoted in Williamson, Radford & Bennetts 2003: 107).

In considering these concerns, new tools such as life-cycle assessment deciding on the basis of what is sustainable, what is preferable and looked-for, and what is economic, will need to be employed. Locally sourced, rather traditional materials and techniques may provide advantageous dynamics, but so can well thought-out combinations with modern materials or technologies. As the cases collected in the Rocky Mountain Institute's "Green Development: Integrating Ecology and Real Estate" (1998), for example, show, such a still rather unusual approach can in fact provide multiple benefits, including economic ones; the novel approaches suggested for the provision of energy are cases for sensible combinations, as well: In all too many cases, better (more modern, and older, respectively) architectural solutions would provide passive/natural heating or, especially, cooling at lower cost than if provided through appliances (certainly once externalities are counted in).

Photovoltaic energy is advancing quickly but still hardly viable in regions where the energy grid is closely knit – although the black-outs during the course of 2003 show that more resilient systems as could be achieved through redundant, local-scale energy supply from simple renewable/inexhaustible sources were worth considering for the sake of security, as well as when thinking towards the future. In marginal areas and developing countries, however, their high reliability and low (to no) operating costs make them the most viable solution already. Within industrialized countries, providers of outdoors equipment such as Brunton are at the forefront of alternative, e.g. solar, energy products for off-grid uses, as they already show their potential in such off-grid situations as expedition and backpacking use as well.

Design for sustainability need not only be an issue for architects, even if the influence of the building's very construction is a great one of course. Like the relation between environmental impact, clothing and personal style, many of its aspects are open to individual agency as well (cp. Seo 2001, for example).

The practical and psychological relations that play a role in the use of plants in enhancing the living conditions provided by the interior built environment draw attention to an aspect of human "integratedness" with nature that is of further importance to an assessment of the suggestions for achieving ways of life more conducive to sustainability.

Human Natural Habitats

Individual buildings are somewhat more easily shaped by individual people's decisions or, for example, the setting of preferable examples which shape expectations and may contribute to improvements (even if following them without considering local ecological relations may not in itself produce higher contributions to sustainability; closer assessment is needed). The consideration of aggregate levels of constructed environments – communities (or at least aggregates of people in particular places), settlement patterns and wider environmental relations – introduces new levels of complication through the "ecology of agents" (Evans 2002a: 22f.) of people and institutions with different capabilities and agendas that become (still more) active at this level.

Yet, they may provide a way of recovering agency (as power to be of influence) in working towards enhanced livelihood and sustainability, as Evans (2002a) argues were the case – not necessarily, but possibly – by community action (cf. Bernard, Young and Jackson, 1996, for example). This even applies in cities, which are particularly strange entities in the quest for sustainability: economic "growth machines" and ecological sinks for material and energy in which many problems are concentrated, on the one hand, as well as centers in which struggles for livability and sustainability find particularly interesting expression in the interactions between the political, economic, and local/community-based (Evans 2002); on the other hand, places the diversity of actors of which could possibly contribute to their function as centers of innovation in both idea, politics, and technology that may be capable of being a boon to sustainability (cp. Moavenzadeh, Hanaki & Baccini 2002).

Community, human settlement, and landscape development, like classical architectural design, would need to be informed by a new mindset oriented primarily on the question of their sustainability (cp. Benson & Roe 2000), with sociocultural concerns playing a high role but needing to be considered in conjunction. Regional planning connected by the image and reality of the landscape, for example, may be an orientation that provides a natural "integrating framework" for connecting "people and place ... meshing ... people's stories ..., indigenously based economic wealth, and sensitive use of natural resources," or "livability ... biodiversity ... prosperity" (Selman 2000: 108f.). Again, such a perspective appears all too radically environmentalist, but in their internal and external ecological relations these concerns could very well provide a positive impetus to both ecological and human aspects.

For one, there are the suggestions for how settled areas could be of reduced impact in terms of their ecological footprint as well as better places to live. These start out with the perspective on "green buildings," but need to be expanded towards the larger scale. Settlement patterns oriented towards people (e.g. rather than cars), together with the promotion of mixed-use of local areas/communities, even including community gardens or straightforward urban or community-supported agriculture, and in general aiming for a better balance of places of living, to meet, and to work, within walking distance or at least connected by public transportation would play a great role. In various instances, such ideas are being tried out, e.g. in eco-villages (cp. http://gen.ecovillage.org), and such struggles for balances between different aspects of urban livability can be argued to have made up a main pattern of, e.g. New York's (cf. Gandy 2003), city history.

Secondly, human beings, like all animals, exhibit particular habitat preferences. Although much research has focused on evidence justifying the view that a preference for savanna-like environments were a genetic predisposition of the human animal, the general gist of such studies is that the promotion of green spaces – with approaches ranging from the savanna/park-like via that fitting regional landscape character, to the wilder areas – can be of great benefit for well-being (cp. Heerwagen & Orians 1993; which will be discussed in later chapters, and especially in Part IV) accompanying and beyond their utilitarian contributions to essential ecosystem services. Kellert (1996: 190–194; 1997) strongly argues for the positive potential that a stronger concern for urban nature, of "captur[ing] the aesthetic and ecological virtues of the natural environment and weav[ing] them into the lives of urban families, neighborhoods, and the places where people work" (1996: 192) would provide. Research supports this view with hard facts: Stress, for example, is an increasing problem in Western societies. Its mitigation can be significantly affected simply by the occurrence of nearby open, green, spaces or gardens (Grahn & Stigsdotter 2003; cp. Ch. 16).

An urban intensification connecting such approaches could, by that very combination, make cities have less of a negative environmental impact, as well as (re)vitalized by promoting human needs for their (our) "habitats" and social structuring (cp. O'Meara 1999). At the same time, such development could be instrumental in reducing sprawl, contributing to both classical conservation (of areas less modified by human influence), and/or to biodiversity (and ecological functioning) even within more densely settled areas, which can be surprisingly prominent features of cities.

The latter idea leads to Rosenzweig's (2003) suggestion of "reconciliation ecology" oriented on expanding the design ideas for green spaces to an active provision of (natural) habitats for many species "in the midst of human enterprise." Considering (as he does) how many areas are now under human influence, classical conservation by setting aside wild/natural, i.e. relatively little-impacted areas will not be sufficient. With reconciliation, including restoration where helpful, e.g. in providing for both wetland habitats and better water quality, vegetation cover and erosion prevention, and oriented more towards rural areas with eco-agriculture and its work towards reconciliation/integration, the earlier could provide (natural, species and/or ecosystem) core areas, the latter corridors to link them.

There appear to be two major obstacles, both by and large approachable with some sort of education: On the one hand, there is a misunderstanding of the ways in which rather wild areas with their own processes could be upheld and could contribute to positive human experience (see Part IV) in a balanced way, e.g. exerting some influence where necessary, but not falling into the trap of wanting to control everything, nor of seeing nature as ideological ideal rather than actor of its own –

predators, for example, need to be "handled with care," prevented from becoming too unafraid of human beings.

On the other hand, there is a need to tackle the assumption (that is also related to human estrangement from nature) that sustainability/conservation were only about some far-away charismatic megafauna when it is just as much about local ecological functioning and species.

Ultimately, there are a number of utilitarian reasons for conserving biodiversity and ecological functioning around, in (and eventually, of) cities, but how well we take care of "our own backyards" will also be indicative of how well we can fare with regard to "our global environmental fate" (Murphy 1986: 76).

Diversity, once again, is an essential part of the consideration: Cities in particular are hotbeds of creativity, as well as of stress, celebrated by those who can stand and even like the advantages of "the heat of the kitchen" (Hall 1998: 989) for those advantages. There are instances of age-old urban ways of life that have been less changed essentially, on the other hand (even the central role of conurbations as market centers, for example, has been among its roles probably since the first invention of the city); and other people less inclined towards crowdedness and stress will prefer other types of settlements and lifestyles. Even though cities achieve their "golden ages" by ("high") cultural creativity, technological innovation, and the combination of those two, and always have to struggle with "the urban order," issues that fall under the rubric of sustainability will be major factors affecting even their future (cp. Hall 1998).

Further relations contributing to a "positive ecology" of sensibly (re-)engaging in contact with nature still remain to be considered, but the lessons of orientations on sustainability for wider concerns of security, rather than just those of immediate physical safety which the construction of environments is a response to, yet require another closer look at the interactions between ecological processes and human concerns.

Chapter 14

Security and Sustainability

The more immediate need for physical safety is mainly satisfied, together with part of the need for temperature regulation, by the construction of environments. Even the relationship between temperature regulation, physical safety, and the constructs used to their satisfaction stands in a precarious balance between technology, ordinary and extraordinary conditions (such as normal environmental circumstances and changes, and natural catastrophes, respectively), and human psychology.

The need for safety also finds expression at the wider scale of a need for – a sense of – security. And here, too, feelings and objective measures can greatly differ, depending both on actual circumstances and on awareness of different factors of danger. Moreover, they vary with dominant psychological mechanisms; the issues taking precedence may reflect or hide the dynamics that are empirically effective in any given situation (cp. Glassner 2000, for example). – Whereas absolute safety cannot be had (even if psychological patterns might prefer for it to be the case), it is possible to show which situations and further developments provide more supportive dynamics than others.

Recently, just at a time when concerns over the possibility of (nuclear, large-scale) war were abating, the "9/11" attacks on the WTC and the Pentagon brought a traditional, warfare-related, concern over security to the fore yet again. At the same time (not least being followed by the Anthrax scare), it underscored some of the changes that have recently developed: capabilities for wreaking havoc are increasing with the spread of global venues for information and communication, "dual-use" (constructive or destructive, that is) technologies, and, on the other hand, modern societies' dependence on high technology as well as the concentration of people and resources, which make asymmetric kinds of warfare more effective.

Although justified in some regards, the focus that the "war on terror" has drawn to conventional terms of conceptualizing (threats to) security is diverting attention from ecological and cultural relations which are important factors shaping the present and future security situation as well, and likely becoming even more so, e.g. those of resources, ecosystem services, global change, and intercultural/developmental aspects (cp. Myers & Myers 1993, Renner 1996, Pirages & DeGeest 2004).

In long-term perspective, access to resources may once again become a cause for international as well as intra-national conflict. As is relatively well realized, this applies specifically (already, many would say) to oil reserves, but rather more importantly could encompass conflict over water and other natural capital (cp. Klare 2002).

Relationships between resource availability/control and conflict, at regional and intra-national scales arise not only from struggle over access to resources but also from the differential accrual of environmental costs and benefits to different segments of societies; even less favorably, the relationship sometimes takes the form of natural resource exploitation financing warfare, as well (Renner 2002).

Another, related and yet more ecology-dependent perspective on security arises from a consideration of local ecosystem services and/or character: Natural disasters regularly displace people at least temporarily and cause increasing damages. Some, e.g. extremes of weather, are definitely adamant to human attempts at control (but likely to be exacerbated by climate change), but their effects could be mitigated by both technological and behavioral "adaptation." In many cases where ecological dynamics play a role, e.g. forest fires, flooding, droughts, both more research into these patterns and action based on the best of such knowledge would be a necessity. In spite of disastrous floods, houses are still being built in areas which are or used to be floodplains (and not in ways so that they could withstand their effects); dams can work for a while, but in unforeseen extremes or over time an ecological solution working with catchment ecosystems, floodplains, etc. is more likely to function sustainably while providing additional benefits such as water cleaning and wildlife habitat. Ultimately, the human-induced/influenced switch of ecosystem states to degraded ones, which (considering the archaeological evidence) is a recurrent pattern of human history, could compel more permanent and large-scale migrations. Already, there are a number of "environmental refugees," and their number is expected by the UN to increase to 20 percent of the world population in 2020 (George 1993).

Going the step further to the new ecological/ecosphere perspective of global change, the potential effects are naturally even greater. Climate change is the most prominent contemporary example, but other changes in ecosystem character and service – which would go hand-in-hand with and make up much of the impact of climate change – could grow to global effect as well. Many scenarios of global change may be far from probable, currently, but are just as far from impossible; the reverberations would be all the greater. One scenario reportedly produced for the Pentagon is particularly telling for its explicit focus on the global political/security situation that a collapse of the thermohaline circulation in the North Atlantic could produce (Schwartz & Randall 2003; cf. David Stipp: "Climate Collapse. The Pentagon's Weather Nightmare," Fortune, January 26, 2004; the scientific basis can be found in Clark, P.U. et al. 2002, for example).

Finally, a consideration of cultural interrelations between the above aspects of "ecological security" requiring a focus on sustainability, and the focus of "positive ecology" on well-being and development, informs the discussion well.

As action in the environmentalist/sustainability arena in general is negatively affected by feelings of futility and lack of visions for positive progress, so life chances and perceptions of them are among the factors that influence what tools of advancement and self-efficacy are being used. When, as with the "American Dream" or the current situation in (East) China, life chances are seen to be great and contingent on individual education and achievement, those are the methods that may be employed. Clashes between modernization and tradition under circumstances of unequal power relations, contributing to impoverishment and/or at least the perception of neo-colonialism, however, are creating a pool of disenfranchised people more likely to resort to politics of violence including terrorism (Pirages & DeGeest 2004: 218). Charles Rojzman (2002), for one, has been arguing that similar mechanisms of unequal power and perceptions of futility, lack of self-efficacy and life chances, are at work in the increase of violence even in European cities (as violent behavior is among the few possibilities for being active and thus gaining some feeling of control always left to people, particularly those who do not see other chances).

Migration patterns, too, already appear to be based not only on actual living conditions but also on the perception of differential life chances, e.g. own (perceived) poverty as compared to the life situation of other countries' inhabitants – as seen on TV. Neither the positive sides of much of what makes up development, nor the factors that make migrants leave their countries should be denigrated. Yet, it is telling to see that Norberg-Hodge's International Society for Ecology and Culture (ISEC) is, to the author's knowledge, the only institution giving (in this case Ladakhi) "poor" people a chance to compare their situation to that actually found in the "golden North," to discover that not everybody there is rich, either (with a "counter-development" book and "reality tours," for example, cf. www.isec.org.uk).

Globalization, which has been changing the security situation in a number of ways (cp. Pirages & DeGeest 2004), is the necessary context in which to consider the above, and especially the ecological and cultural, factors: Distance, enforceable by military measures, was seen to provide protection from far-away happenings, but with global change(s) and globalization, it cannot do so anymore. – As Pirages and DeGeest (2004: 218) note, "in the emerging global city people move rapidly from one neighborhood to another; containment is not a viable option." (So, of course, do microbes, plant and animal species move.)

In contrast, global governance, orientations on sustainability in its environmental/ecological aspects as well as in its positive contributions to fulfillment of needs, living conditions, and well-being, and including in concerns with social justice, intra- and intergenerational equity, then become essential mechanisms by which to work towards security. Reinforcing security by military force will likely remain a concern (and probably an unfortunate necessity) of the state. However, taking the form of talking softly and carrying a big stick, as Theodore Roosevelt suggested, and especially accompanied by measures towards global governance, equity and sustainability, it is more likely to provide the intended effect. By force alone (or even in conjunction with security technology, unless willing to turn from an open and globalized society to a newly segregated, illiberal world), it will not provide more than a false sense of security. – In the argument for the necessity of peacekeeping activities, at least, this seems to be common knowledge; neither does it take a rocket scientist to come to such a conclusion (cp. Lessig 2004, for example).

One potentially helpful input from traditional approaches to security should still be mentioned, which is the lesson that Prince Charles mentioned in a statement quoted in "Our Final Hour" (Rees 2003: 112f., where it was incidentally put in the context of how even a small chance of runaway climatic change would justify drastic measures to prevent it): "The strategic threats posed by global environment and development problems are the most complex, interwoven and potentially devastating of all the challenges to our security. .... In military affairs, policy has long been based on the dictum that we should be prepared for the worst case. Why should it be so different when the security is that of the planet and our long-term future?"

The challenge, however, is that the current orientation is not even based on preparation for the worst case, but only on hope for the best in both what change may occur and how societies could adapt. The alternative possibility of opting for prevention and a focus on sustainability, although containing potential for the better, is disregarded for the break with what is considered a normal – industrial, profit-oriented-only – course of affairs it would entail. Munich Re, in its "Annual Review of Natural Catastrophes 2003," too, sees its risk researchers' analogous predictions increasingly confirmed: "1. The risk of pronounced and prolonged periods of heat and drought is increasing dramatically in large parts of Europe [and globally]. 2. The potential effects on the economy and the insurance industry are completely underestimated or even ignored. 3. There has been a ruinous neglect of precautions for adaptation to such situations" (Munich Re 2004: 22).

Only in one prominent instance has the argument for adaptation to climate change recently emerged: "The defenders of business-as-usual on climate change" had first been arguing for the non-existence of any such problem, then accepted it and argued that there were enough time to do something, and finally – recently – concluded that it were too late but to wait and see what will come of it and try to adapt; and "remarkably, the Bush administration moved through this string of evasions in half a presidential term" (Speth 2004: 6).

Looking back to potential scenarios such as that for the Pentagon, and onto history, the need to consider precaution should be obvious – even in historic times, and outside of human-induced global change, extreme climatic events such as the European medieval (Baroque) "Little Ice Age" or the American "Dust Bowl," have occurred.

Sustainability-oriented alternatives, taking their cue from evolutionary-ecological theory (cp. Pirages & DeGeest 2004), would likely provide further positive contributions towards security (and in other relations). In such an approach, resilience needs to be a major focus, after all. In many modern circumstances, e.g. with IT (where some such systems are quite commonplace) or with the provision of energy, such approaches oriented on resilience by redundancy and low input requirements, e.g. through local sufficiency, long life span of solutions, little maintenance (other than locally available), and integration into or with larger-area networks for backup, can be both better for sustainability, as well as more secure.

In the case of energy, such suggestions have most recently come from the Rocky Mountain Institute's publication "Small is Profitable" (Lovins et al. 2002), which is not even focused on security but on economic benefit. Weather extremes, incidentally, are among the most instructive occurrences. In the summer heat of 2003, water and nuclear energy production facilities experienced problems because of general lack of water and too high a temperature of the water used for cooling, respectively, while the demand for electricity (for air conditioning) surged – earlier, solar technology was disregarded because the peak of solar radiation came in summer when least energy was supposedly needed. In winter conditions, on the other hand, power lines can sometimes get toppled by ice storms, leaving households and entire cities without electricity, while local, redundant systems could at least contain the impact.

Similar orientations, as were suggested in Chapter 12, can make the food system more sustainable as well as more secure, by a de- and re-integration from distorted (ecologically and economically subsidized) markets, and with local eco-cultural conditions and resilience and a global trade and/or buffer stock system, respectively. Further positive effects, e.g. on health, are likely as they counter the spreading epidemic of obesity.

To draw on the work of Pirages and DeGeest for a last word, there are two different paths towards sustainability: One by reactively muddling to it and being forced in such a direction through crises and violence, the other to work towards it by "anticipatory thinking, initiative, creativity, and cooperative engagement to forge agreements and frame policies that can mitigate much of the potential pain involved in moving toward a less materially voracious, but much more egalitarian and ecologically sustainable world" (2004: 206f.).

Chapter 15

Health

Health is an issue rather like security: highly relevant to the individual but meaningfully analyzable almost exclusively as a statistical measure (of population health), and becoming an issue mainly just when its absence makes it conspicuously so. In yet another similarity, health is intimately related to, or even an essential element of, sustainability in its focus on human survival and well-being. Yet, ecological interactions with health – "health ecology" (Honari & Boleyn 1999), even – do not hold a prominent position. In particular, a wider view on health not as a kind of commodity but as a condition determined by interactions between ecology, society, and technology, and in long-term perspective contingent on sustainability (at the least, of the natural life-support systems) is of necessity (cp. McMichael 2002a, 2001).

Diseases, not least through their obvious, negative 'naturalness' and through recent developments of globalization (such as between SARS spread/scare and air travel), are one major factor in considerations of health, sustainability and security. However, the relationship between population, people and disease is one of the most obvious contemporary examples of how difficult it is to suggest appropriate balances between humanity and nature (including a necessary element of "biophobia"), and how misleading the separation suggested by such terminologies is (since the human position as and within nature cannot be escaped).

The long-standing approach of a war to eradicate diseases through medicinal technology has had some success (with polio and smallpox), but both (twenty) diseases thought to have been coming under control, e.g. tuberculosis, malaria, cholera, have made a comeback, and (thirty) new diseases, particularly HIV/AIDS, been found since 1973 (Pirages & DeGeest 2004: 149). Meanwhile "bioinsecurity" (ibid.) is growing still further, and doing so through changes in human population dynamics (densities in urbanization) and behavior (incursion into previously uninhabited areas, increasing travel and transport, etc.), as well as directly through environmental degradation. Obviously, these issues are strongly related to concerns of sustainability.

The orientation which is to guide the future approach to health – as to security and the other concerns included in the heading sustainability – is highly relevant in this context. However, as in assessments of security, metaphors that may not be appropriate still guide too many decisions: Attempts at eradication of microbes by having even household cleaners have disinfectant effects, let alone the use of antibiotics as growth stimulus rather than against disease in feedlot animal farming, for example, have been contributing to the resurgence of diseases, in these cases by promoting the build-up of resistance, because the evolutionary dynamics of pathogens have not been brought into the consideration (cp. WHO 2001). Yet, the description of human health as a battleground between diseases and the body continues to motivate the debate, leading to the hunt for new "silver bullet"-solutions for the eradication of pathogens.

Under conditions where health is less threatened, however, some exposure to microbes that are less or not dangerous, to endotoxins, etc., is in fact, to the contrary of popular images, necessary. Both the development and the functioning of the organism (Dusheck 2002: 57, Rawls et al. 2004, Hooper & Gordon 2001), requires healthy amounts of (contact with) them. This applies in general, and particularly in regard to the digestive and immune systems (for the latter of which, lack of "training" has been suggested as a major factor contributing to the increasing occurrence of allergies; cf. Kabesch & Lauener 2004, Braun-Fahrlander et al. 2002, Riedler et al. 2001, Yazdanbakhsh 2002). This relationship mainly comes to the fore in a non-threatening way only when the other basic needs described so far, especially nutrition and safe water and hygiene are fulfilled, so that adequate bodily defenses against pathogens are built up and prevention of disease is undertaken, of course.

The view that these contrasting perspectives give rise to is quite different from that of an all-out battle: some diseases need to be avoided, but apparently cannot be eradicated, and life in spite of the existence of many diseases – especially working on preventing their contraction – is quite possible.

It is necessary, then, to consider "sustainable health" as standing in balance between the need to "battle" diseases, and the impossibility of escaping any and all health problems. – Balance, not war, is the necessary metaphor (cp. McMichael 2002b). "Evolutionary medicine," as an approach focused on "living with" diseases but in relative health may therefore be worth a consideration. At least, it could help understand the evolution of drug resistance and find ways to prolong the usefulness of new drugs (Stearns 1999; Trevathan et al. 1999), but possibly also in developing a truce between contraction and negative effect of diseases (Ewald 1994, 1999a, b).

Immunization, prevention through education (e.g. on STDs and condoms), together with modern medicine where it is best, do of course have a role to play. As yet, the fulfillment of basic needs and adequate, accessible health care systems are suggested as the major concerns for developing countries (Mascie-Taylor & Karim 2003).

The paramount significance not just of disease prevention per se but of promotion of good health yet needs to be recognized more strongly: Both aspects of health would be contingent on the provision of adequate nutrition, safe water, hygiene, and other such basics, as well as the prevention of pollution, continuing functioning of ecosystem services, etc. The positive effects of these measures would accrue to the population at large, contributing to health in the most widely effective manner. – So, the health situation of the greatest number of people could be improved by relatively simple and less expensive measures focused on issues which are among the goals of sustainable development, anyway.

Water, for example, is a major element of danger to health when functioning as a vector of disease and pollution. With the purification provided as ecosystem service (which is highly effective in itself and can be technologically utilized or improved) and prevention of pollution e.g. by ecological sanitation, "integrated" technology, etc. one major part of water safety could be addressed. Additional technological measures to make water microbiologically safe to drink can then be implemented more easily and cheaply, e.g. by solar water disinfection or filtering, than if all the cleaning had to be achieved by technology only (cp. Ch. 11).

In fact, such approaches not directly focused on the disease but on conditions promoting health could be all the more promising if it was not so much medicine as nutrition and sanitation which improved the state of health in "developed" countries to the largest extent, which is rather likely.

For these reasons already, conservation and practical environmentalism/integration are essential elements of approaches to "sustainable health;" their importance is still greater considering the (potential and actual) significance of traditional medicine which relies heavily on herbal drugs, and considering concerns that are, in the Western view, not directly medicinal:

For one, a large part of the human population is dependent on traditional medicine and would not be able to afford otherwise; a WHO estimate is "that 80% of the people in developing countries of the world rely on traditional medicine for their primary health care needs, and about 85% of traditional medicine involves the use of plant extracts" (Farnsworth 1988). The potential importance of traditional medical knowledge and medicines, and with it the biodiversity utilized, extends to modern science as well, e.g. for the discovery of "new" pharmaceuticals (ibid.).

Just as in the case of traditional crop species and varieties, the situation is highly affected by issues of valuation, intellectual heritage commodification, and just compensation (cp. Moran 1999), as well as by its interaction with more direct socio-economic dynamics (see Lagrotteria & Affolter 1999, for example). Issues of conservation and sustainable harvesting also apply to medicinal plants, of course. For example, cultivation of medicinal herbs can present basically the same problems as any other agriculture – but also the same benefits of eco-agriculture. This may apply even more strongly, with an eye towards conservation, because of the higher esteem, price (and medicinal value) of herbal drugs from the wild.

Secondly, for many of the conditions afflicting health that do not require surgery, prostheses, and other treatments for which modern medicine is the only or best medical system, traditional and traditionally-based remedies and approaches exist and offer distinct advantages. Malaria, for example, is increasingly resistant to synthetic drugs, while effective herbal remedies can still be found and (what is particularly significant) inexpensively produced: Artemisia annua extracts, even in the simple preparation as tea, appear to be effective against it, and this drug can be grown by the people needing it themselves or on farms (the plant usually qualifies as a weed, pointing to the lack of difficulty it presents to cultivation). And quite possibly, the whole plant extract is more potent than the single extract of artemisinin (the main active compound) as it contains further substances contributing to the anti-malarial effect (cp. Wright 2002 and the review by Duke 2003; Mueller et al. 2000).

The best outlook on health, which is promoted both in its definition by the WHO and in traditional medical views, however, is one that is focused not so much on the absence of disease, but on good health and even on well-being – which in containing an overall state of good health does, of course, imply better resistance to diseases and/or their ill effects. Yet, well-being also implies being able to live with the problems that life naturally brings with itself.

The importance of prevention by healthy ways of life needs to be emphasized in this context. Its foundations still lie with basic needs, e.g. sanitation, adequate nutrition, basic health care. In summarizing that these are among the most pressing needs for developing countries, Mascie-Taylor & Karim (2003) suggest that healthier lifestyles and public health education are, in fact, the most necessary measures in industrialized countries as well. Considering the great effect that, on the one hand, education about health problems and their avoidance (witness the continuing spread of HIV/AIDS), and healthy ways of life such as ones that would reduce the global obesity epidemic and the infirmities associated with it, on the other hand, could have on population health, the importance of these aspects of individual lifestyle and culture (as habitual ways of life) equally apply to developing countries.

The potential contribution of food to good health, for example, interrelates nutrition, health and (or as) well-being even more strongly than was done in Chapter 12:

The fundamental basis lies in the balance of caloric intake and exertion, as well as in the meeting of nutritional requirements of both macro- and micronutrients and secondary phytochemicals. Apparently, this is best achieved by "poor" diets that are calorie-reduced (cp. Anson et al. 2003), relatively low on animal protein, high on staples, vegetables, and fruit. These relationships between ways of life and food ways among indigenous peoples (and traditional societies) included a balance with the particular properties of the consumed foods, and with physiological and cultural-behavioral adaptations (cp. Nabhan 2004). Examples of these relations are given by the high risk of Polynesians and Amerindians of suffering from non-insulin dependent diabetes mellitus and obesity under conditions of a shift to a "modern" diet (cp. Young 1993), and the benefits of traditional diets which were shown to occur among Australian aborigines (O'dea 1984) and Hawaiians (Shintani et al. 1991), at least (referenced in Johns 1999: 163).

Not surprisingly, traditional medicinal systems such as Ayurveda or traditional Chinese medicine emphasize the healthful aspects of certain foods and food combinations. In such regards, here it is important to note that organically produced foods quite probably have a better effect on health, not only by not containing pesticides, but by containing more secondary phytochemicals – the actual relations are not quite clear, but scientific studies offer promising suggestions that the effects of less-processed ("whole") foods from organic agriculture (thereby) promote health and well-being (Soil Association 2001, Velimirov & Mueller 2004, Cleeton 2004, Huber & Fuchs 2002).

It does not take scientific studies (or ideological diet plants) to realize such potential: Sports nutrition consultants (and the USDA's) recommendations for active people, featured in Outside Magazine, pointed out the positive contributions of a diet that is diverse and "whole foods"-oriented. Rich in fruits and vegetables, and consequently in a variety of both necessary macro- and micronutrients, such a diet is not only good for energy (nutrition) but also for health and performance of athletes and outdoors people (and everybody; cf. Hewitt 2003). Last, but not least, feeding trials with different animals have shown recurring preference for organic produce as well as higher fertility when fed with organically grown plants (see Velimirov & Mueller 2004).

The content in particular phytochemicals (usually always the one most recently discovered and most prominent in the media) is a common theme of marketing for enhanced "functional" foods, of course. The best source, however, are traditional combinations and natural diets; the reliance of modern nutrition on a very limited number of species and varieties may actually be too restricted to provide many useful/healthful compounds which traditional diets would contain (cf. Bogin 1991).

Foods do not only form the basis for good health through the important health effects of both malnutrition and overconsumption, but exhibit medicinal side effects. For example, "vitamins and phytochemicals can have positive effects on immune function or may have antibiotic effects that are important in controlling parasites and infectious disease" (Johns 1999, referencing Johns 1990).

Eventually, traditional diets contain some herbs and food combinations with genuinely medicinal effect: "medicinal foods" and "wild" plants in the Hausa (Etkin & Ross 1994) and Maasai (Chapman et al. 1997) diets, in Polynesia (Cox 1994), in Amazonia (Dufour & Wilson 1994, Vickers 1994), and in North American indigenous peoples' diets (Moerman 1994). Less exotically, but only as a 'survival' of the diversity of food combinations for medicinal purposes, such a relationship may have been the (scientifically ultimate) reason for the widespread combinations of foods with spices (the proximate reason being that it tastes well; cp. Sherman & Flaxman 2001, Sherman & Billing 1999).

Interestingly, the epidemic of obesity in its need for countermeasures consisting of education and/or culture change as well as policy, now suffers from the same problem as a transformation to sustainability: Getting across some message works relatively well, but it is soon co-opted for marketing; by and large, little practical change is achieved. Meanwhile, the problems exacerbate because, rather than learning from the problems, "developing" countries have started to follow the exact same path.

Quite a different problem than with natural pathogens is presented by novel substances, their spread, accumulation and effects. The need to avoid their production (or at least release) as an aspect of an economy/technology for sustainability has been mentioned before. It not only refers to ecological considerations, however, but also to ones of health: hormonal (endocrine) effects, physiological changes (and potential psychological effects as pollutants can impair mental development), aggravated by the potential effects of long-term exposure and exposure to mixtures of novel substances (which are only tested for toxicity as single substances), etc. apply to human beings all the same (cp. Schettler et al. 2000, Myers et al. 2004, www.ourstolenfuture.org), and as – ecologically speaking – "top-level predators/omnivores," accumulation in food webs means that any substance that can get enriched in living tissue will end up on our plates and in us.

So, human ecology and management of ecological and social systems as well as culture, e.g. in terms of food and lifestyle, will, as it has throughout history, "shape the patterns of population health and survival" in the future (McMichael 2002c: 1147).

Ultimately, the relation between sustainability and well-being suggested as the core for "positive ecology" comes to the fore most strongly when basic needs including health are satisfied, and the relation in a "higher scale" cultural integratedness with, and psychological relation to, ecology becomes effective. These aspects and interactions will be addressed in the following part.

PART 4
Eco-Cultural Integration

Ecology in Culture and "the Good Life"

Nature and culture (whether cognition, behavior, or both) have commonly been considered independent since the departure from theories of determinism, even in recent times where biologism in the form of genetic determinism apparently tends to become rampant again. Yet, even culture, knowledge, and spirituality are, to a large extent, intimately bound up in relationships with the environment and human experiential engagement with(in) it, as are their expressions in the individual and his/her psychological well-being.

More specialized human, cultural forms of social, psychological and cognitive patterns, most notably spirituality, depart from mere survival and represent an actualization of a second, non-biological system of (cultural) evolution transmitting its elements by learning. As a matter of course, it interacts with biology and evolution, at the very least by being based on the organism, its structures, capacities, and possibly predispositions, and (basic) needs that have been developing in biological evolution, and vice versa by shaping human action including, for example, modification of the environment.

To considerable extent, such elements of human life nevertheless are basic needs having to be addressed even for mere survival, as well. – Social relations, sensible management of the information derived from perceiving the world, and even a sense of self-control over life (self-efficacy) all take forms in the human species and its diverse cultures that seem unique and are indeed special, but are necessary for any species' survival. Finding a strong relation between biological and cultural diversity in any area of life among indigenous peoples and in subsistence economies is hardly surprising, as such lifestyles are intimately bound up in ecological workings, e.g. with the cycle of the seasons. The interrelation is seen as strong enough to justify calling it "biocultural diversity" (cp. Maffi 2001). Even in modern situations and societies, and in regard to the "higher-scale," e.g. cultural, psychological aspects of life, however, mutual relationships apparently – as will be shown in the course of the following chapters – continue to exert their influence; nature is valued not only for strictly utilitarian reasons.

Both considerations – relations between ecology, human biology, culture and well-being, their necessity, history, and potential, and on the other hand the farther independence from ecology and impact on it – need to play a role in considering sustainability. On the one hand, certain forms of or features in culture(s) may be more amenable to human well-being in light of our biological foundations and psychology, e.g. as developed during the course of evolution; on the other hand, culture may be progressive (as opposed to biological evolution, and as measured against an ideal that is itself cultural), but need not be adaptive in any and all instances. Rather, much of its evolution/development follows what Durham (1991) has called "secondary value selection," i.e. is based not on its value for survival and/or impact on reproductive success (natural and sexual selection), but rather on "fit(ted)ness" in terms of other cultural features. Hence, as is most obvious in the modern industrial culture, not all of culture needs to be positively contributing to human well-being, let alone to sustainability. Progress, learning from past mistakes and future scenarios, would nevertheless be possible – and in this context, highly recommendable.

The essential and "higher-scale" character of such concerns – Maslow's "higher-order needs" (1970) – informs a complementary perspective to the more basic needs and their (usually more direct and material, hence more apparent) relation to the ecosphere:

The relationship between culture and nature may be much more obvious in indigenous societies, but the relationship still continues to exert its pull on modern humans (not surprisingly, since our modern time is yet but a blink of an eye in our evolutionary history). The possibilities of how these relationships can contribute to human well-being are far-ranging, and provide a yet stronger perspective to (a) positive ecology. Hope for the future, possibilities for progress and individual well-being and happiness, after all, are just about as important motivations as basic needs are (in part, as just argued, they are additional needs themselves). And although their relation to a biodiverse environment – apart from its role as foundation without which everything else were moot, which is apparently not strong enough an argument to achieve widespread change – is a difficult issue, particularly when it competes with the easy promises of happiness of consumerism, it is of effect and of considerable interest to many people. Individuals and initiatives around the world have been working from such a point of view that different visions of a good life are necessary, e.g. a "New American Dream" (cf. www.newdream.org). It is necessary at least to consider the question that Kellert (1996: 32) raises: "People can survive the extirpation of many life forms just as they may endure polluted water, fouled air, and contaminated soils. But will this impoverished condition permit people to prosper physically, emotionally, intellectually, and spiritually?"

From a scientific perspective, too, a considerable number of human valuations of nature – humanistic, symbolic, naturalistic, moralistic, and ecologistic-scientific ones (cp. Kellert 1993, 1996, 1997) – come to the fore in the context that is now to be discussed: In what further ways sustainability can be shown to be a basis for a positive ecology in which it is not only a necessity for human survival, but also offering great chances for living a fulfilling life, pursuing personal satisfaction and happiness in ways that do not endanger other people's and future generation's possibilities for doing so.

It is also an interesting feature of such relations that their contribution to well-being is more contingent on knowledge and openness to the experience than on money. Outdoors experiences, for example, can strongly profit from learning new knowledge or skills. They can also improve with technological tools, but do not necessarily require them. Moreover, they can take place in far-away, exotic places, but would quite similarly be possible in a nearby nature. Also, the needs for relaxation, sociality, etc. in their ecological relationships point out that the view on the relationship between (financially) poor and well-off people or societies and nature needs to be expanded yet further: They are not only a luxury for the rich, not just a necessity for the poor's survival, but nature in its linkage with needs is both necessity and luxury, i.e. fundamental to human survival and a possibility for enhanced well-being to both rich and poor.

Chapter 16

Rest and Arousal

Sleep, obviously, is a very basic human need; so are rest/relaxation and its opposite twin of arousal. Less obvious are their relations to culture, on the one hand, and to human ecology and therefore sustainability, on the other hand. Then again, culture in the popular sense of arts, "high cultural" entertainment would even be equated with particular ways of arousal. Although this is not the way in which "culture" is used in anthropology (cf. Harris 1999) – or here, usually – those diversions do form part of the concept in its inclusive sense. Fascinatingly enough, these same entertaining, arousing activities can serve as distracting, relaxing pastimes, as well.

All three aspects – sleep, relaxation, and arousal – appear to be quite unrelated to sustainability, or more generally to the environment, however. As long as the connections between environment, person, and culture are overlooked, this may appear to be so, whether it occurs because of their obviousness or because of the lack of support for analyses which consider such linkages across disciplinary boundaries: Sleep relates with conditions of a modern life that are usually taken for granted but may be worth a closer look, questioning their impact on both people and planet; arousal relates entertainment to the environment by way of the impact of consumption caused by it, or even more directly when it takes place in the outdoors; consideration of possibilities for rest/relaxation brings the importance of nearby nature to the fore, and provides linkages with the "higher-scale" elements of needs and/or culture in cognition and creativity.

In this regard, several of the "primal diversions" Wilson (2002: 40) described as distractions from the "real business of sustainability" as much as they are, in their divergent relations to nature/environment and importance to human beings, elements essential to the suggestion of a positive ecology that works to motivate a transformation to sustainability.

"Corpo-reality," as an expression of how humans as living beings are a part of nature is an often overlooked fact that provides one important starting point – as it does in providing the relation between humans and the world through the ingestion of environment in the form of food. The human need for sleep is a very direct biological implication of our being living beings. Once again, the need is interrelated with other factors, e.g. with (a sense of) physical safety as a preferable precondition, and with health as another need that is interrelated with it. Concerns of sustainability such as biological conservation are not directly related to this need, of course, but indirectly they are soon implicated: The widespread attempt at turning night into daytime through artificial lighting, for example, does exert its effect on animal life, whether that be animals navigating by moonlight (which is now drowned out by artificial light) or humans lacking in sleep for having spent the night working or partying. The associated energy consumption further contributes to effects on/against sustainability. Clearly, the extension of usable time through lighting has its positive sides (this work would not have been finished without it), and the critique is not meant to imply that there need not be any nightly partying or work. Following any "progress" that would prefer if humans were capable of working 24/7, however, should be open to questioning.

The approach towards the body as a machine that should function however one, or one's employer, wants it to is all too good an expression of the intolerance to biology that also finds its expression in the incompatibility between nature and human life/living that we have been building up in technology and economics. As this work is suggesting that advantages, (especially immaterial) profits could be had from working with nature, so the relation of human bodily functioning to the environment mediated by sleep raises the suggestion that better regard for circadian rhythms, less nightly noise and energy consumption, i.e. less noise and less work or mobility at nighttime is an idea worth consideration: Chronobiology used in medicine and for wellness, for example, shows possible advantages of such an approach for health; the trouble with disregarding it is shown by the finding that nightshift work is related to health problems (cf. Cervinka 1993, Costa 1996) as well as accidents (cf. Leger 1994, Dinges 1995). Even cognitive functioning is dependent on sufficient sleep (cf. Maquet 2001, Wagner et al. 2004).

The relation between pleasurable experiences of and for rest/relaxation or arousal and sustainability is yet more apparent as the consumption of energy and products related to entertainment, for example, is among the effects of "primal diversions" that would soon be regarded in this context. And, they would be a point in which, seemingly of necessity, the argument for sacrificial altruism would surface.

Once again, however, both considerations of ecology and of anthropology/psychology need to be considered: Technological solutions for making activities such as media consumption be provided with fewer negative effects on the environment, more sustainably, likely could be developed, e.g. with ecological design, resource efficiency/efficacy, etc. The other side of the equation lies with the need for consumption such as for rest or arousal: Clearly, technological solutions are not the only ones available and being pursued – people take pleasure in many kinds of activities, many of which are less consumptive or providing both for needs and for support of more sustainable lifestyles: think of the get-togethers for a diversity of foods and drink (the greatest diversity of which is, as seen above, not found in supermarkets and from industrial agriculture), as of the Slow Foods movement which can simultaneously serve purposes of nutrition, pleasure, arousal and/or relaxation.

Furthermore, many of these activities actually take place in nature, hence do need it, and are capable of being performed in ways that are not, or at least of less negative impact. Possibilities for exploration, for having pleasurable experiences, are well provided by outdoors activities and quite capable of contributing to well-being.

Considering the theory of flow (Czikszentmihaly xxxx), it is likely that the engagement in the activity per se, e.g. physical activities such as runs or hikes, are already of better effect than passive consumption (and they further contribute to good health). Exploring new, interesting or more difficult trails, and increasing skill levels at it, hopefully even gaining competence in skills such as the Leave No Trace approach to hiking would contribute further to positive effects. Particular relevance, however, lies exactly in many outdoors activities' low demand for specialized equipment and skills.

Interestingly, nature experience and outdoors activities are particularly effective in providing a positive contribution to rest/relaxation, but also to restoration, and even for cognitive functioning (for the latter, see Ch. 19). They need not even take place in grandiose landscapes to do so, elements of nature or nature activities tested and found to have positive effect were, for example:

- trees along city streets (Sheets & Manzer 1991)
- neighborhood parks (Ulrich 1981)
- walks in urban parks, even when controlling for the effect of exercise (Hartig, Mang & Evans 1991)
- green views from hospital rooms (which improved recovery from surgery; Ulrich 1984)
- green views in general (tested for their contribution to residents' satisfaction with their neighborhoods and aspects of their sense of well-being; R. Kaplan 2001)
- (videos of) drives in nature- (as compared to artifact-)dominated areas (Parsons et al. 1998; although less recommendable in terms of an orientation on sustainability)
- nature reserves as compared to rooms with tree views, rooms with no view, and urban settings, where positive contributions were overall found to be greater or even occurring only in the former settings, but lesser or even negative effects in the latter (Hartig et al. 2003)
- gardening (Kaplan 1973; which is even used as psychological therapy, cp. American Horticultural Therapy Association at www.ahta.org)

In their review, Ulrich et al. (1991) also concluded that stress mitigation is a recurrent benefit people derive from recreation experiences in nature areas, even urban ones.

Restorative effects do not only occur for adults in cities, "nearby nature" helps even rural children cope with stress, in itself and by promoting social interaction and providing emotional support (Wells & Evans 2003); effects of interaction with nature during children's development are even more interesting because of the suggestion that both values and particular, generally positively connoted skills (e.g. orientation, concentration, autonomy) are significantly improved by it (cf. Kahn & Kellert 2002; Wells 2000); though environmentalist attitudes are not necessarily (cf. Hammit, Bixler & Floyd 2002).

Of course, the occurrence of both positive effects on sustainability/conservation as well as on well-being is a matter of balance. As long as the nature worthy of conservation were only the above nearby elements providing relaxation, for example, the considerations just mentioned would not contribute much to support for conservation in general. This, however, is hardly to be feared. More commonly, far-away wild nature would be the only element recognized as providing great experiences, and possibly seen as being threatened by the existence of people living and (too many) tourists visiting there, but be visited nonetheless. Thus, conservation may be supported, but along with lifestyles hardly conducive to sustainability. Outdoors activities could also take place in nature only to add an element of danger and wildness to activities which are disturbing or outright destructive to this same wilderness; adventure races, for example, tend to use nature as mere setting while the high level of exertion mainly directs attention inwards.

The issues are far from being resolved completely – even the relationships between recreation activities or settings and benefits derived are not well-understood scientifically (Pierskalla et al. 2004); the relations between participation in outdoor recreation and environmental concern, too, are not entirely clear, though pointing to a positive association between the one and the other in general, but not the same across all measures (Teisl & O'Brien 2003).

Anyway, that it should be possible to moderate the effects, whether by changing the technology employed or by coming to see that there very likely is interesting nature worthy of exploration – if one takes the time to learn about and engage with it – even in city parks or backyards without having to lead a bleak existence should become understandable.

The contribution to well-being, too, is standing in a balance between inherent natural/biological relations, and effects of knowledge and experience. Having built up obesity, for example, or not having had a chance to play alone in more natural areas, makes a positive experience of activities in the outdoors rather unlikely. City children, for example, have been shown to prefer the environments and activities (malls, paved roads) they are used to, and to value comfort more highly than rural children who know how and what to do in the outdoors (cf. Hammit, Bixler & Floyd 2002).

Such relations are rather conspicuous in the attitude towards exploration/adventure, as well – the stories one nowadays gets to hear seem to indicate that there were no more adventures to be had, unless by the most extreme of sports and experiences. With knowledge of how and what can be found, and increasing skill at it over time, however, even nature in the backyard, urban parks, etc. likely holds surprising diversity, hence possibilities for ("micro-")adventures.

On the other hand, positive effects of nature experience and moderate physical exertion on physical and psychological functioning, health, and to well-being (as those mentioned above), appear to be straightforward relations based on our biological foundations. These observations are informed by, and contribute to, the perspective of what Kellert (1993) has classified as the "naturalistic" valuation of nature, i.e. "the satisfaction derived from direct contact with nature" facilitating curiosity, mental and physical development and functioning, and outdoor skills.

Hence, they still do need an initial expenditure of psychological energy that is higher than that necessary for just falling on a couch and turning on the TV. As humans are inherently active, curious beings attuned to a life in and with nature, however, the experiences obtained, as well as skills that may then be found to be worth learning (and/or developing as a side-effect) are likely to relate to wider well-being and functioning, e.g. psychomotor skills, concentration, exploration – more so than passive consumption of entertainment. Individual differences probably do exert as much of a role as prior experience and cultural support do, but it could be argued that people who state that they explicitly disliked nature "unknowingly suffer for it in terms of their overall physical and psychological health" (Kahn 1999: 38). Coercion would not remedy this situation, but as these people should have possibilities for retreating from nature, those who do like or want to try that should have possibilities for retreating to nature.

Since the needs of rest and arousal have been used to question lifestyles based on passive acceptance of their conditions and passive consumption, one ultimate related aspect is well worth pointing out: Consumption for (or even as) relaxation or arousal seems, when lifestyles oriented towards sustainability are argued to mean having to lead a bleak existence, to be considered one of those luxuries of life that make it most pleasurable. Increasingly, however, control over personal time –"having time," being able to follow one's own rhythms – is coming to be considered the ultimate luxury; products can simply be bought, time (particularly time well spent) cannot be.

The separation between "making a living" and living, deciding about values of one's life and taking what is offered as supposed needs, should thus be even more worth questioning. Ultimately, then, a deeper look at human psychology will be necessary to consider what makes a life ultimately well-lived and well worth having been lived.

Chapter 17

Social Needs and the Company of Nature

In modern societies, a central aspect of life tends to be the support for individual autonomy in the pursuit of happiness, even over social obligations. This may even contribute to subjective well-being, although its social value depends on the view of the observer (Ahuvia 2002). Nevertheless, some form of sociality is a basic need. Furthermore, the factual dependence on others rather than autonomy from them has increased e.g. with the specialization of economic roles. – In the relation to nature as ultimate source of the goods that provide for the basic needs, this is especially clear as "all but a few eccentrics who subject themselves to 'wilderness training' would quickly perish if deprived of the goods and services provided by their fellows" (Farris 1984: 132, quoted in Atran 2001: 165).

In its most direct relation with the environment, the need for sociality finds an expression in the humanistic valuation of nature, that is in "feelings of deep emotional attachment to individual elements of the natural environment ... usually directed at sentient matter, typically the larger vertebrates" (Kellert 1993: 52). It is a strange relationship as it goes together with a valuation and interpretation of animals that holds them most dear the more they seem to be like us, and/or simply considers them to be "good" in some way, but also tends toward ideological, not necessarily verisimilar views.

Companion animals, in particular, are prone to humanization to the point of being treated like relatives or friends. This relationship shows the deep affinity that humans (can) have for the company of animals, which even delivers benefits for human physical and psychological health (cf. Kellert 1993). The affinity for nature is further reflected, in all its philosophical and practical complexity, in the role that nature (and again, particularly animals) plays in human cognition. How well the relationship describes what is (in nature per se), or what should preferably be (in the relation, for considerations of ethics or sustainability) is rather questionable (cp. Shepard 1993).

The relationship gets even stranger, certainly when ideological positions come into play, since animals can at the same time be liked, considered to be alike us, and yet also be killed for the products they provide.

The true progress of a society, it has been said, may be measurable in how it treats animals. By this count, modern industrial agriculture in which we would not want to give up meat-eating but neither wanted to face the necessity of taking animal lives to eat is nothing short of schizophrenic. A position arguing for absolute non-killing has been propounded at times, most consistently by Jainism, but hardly is a solution everywhere at once. More likely, providing a good life to the animals – out of moral, nature-oriented humanistic considerations, as well as because it should provide the better, healthier meat – yet understanding and standing the fact of their deaths for our living, may be reflective of the understanding of a non-ideological orientation for sustainability. As seen in the chapters on nutrition and health, diets with fewer animal products are better for people and planet. Those who would stand for non-killing should really do so, and those who want their steak may still exhibit concern for animal welfare, the animals' and their own health, and then eat meat knowing where it came from. – At the least, this would entail less estrangement from our dependence on and inclusion in nature than either meat production that is hidden from sight, or the ideological argument that everybody had to follow the exact same rules for the sake of the animals.

A relationship between the human social need and nature that is both similarly deep and similarly exhibitive of interpretations that need not be ecologically verisimilar is provided by the "mutual interlocking between the understanding of natural objects and natural processes, on the one hand, and social institutions and specific behaviour on the other ... go(ing) together to form the prevalent cultural orientations" and the use of natural metaphors "for the interpretation and manipulation of social relations" (Bruun & Kalland 1995: 4).

The potential positive effects for human well-being as well as towards the promotion of sustainability of the sense of sociality and community extended to include nature should still not be underestimated: Historically, localities – land, specific natural features, climate, the produce of the land – together with cuisine, traditions, and history provided a sense of community and to many people still do, or could do, so.

Even spiritually, a sense of community/communion with nature is a recurrent theme of the relation of both human beings and the environment, e.g. of elements of nature as agents in their own right which are yet indivisible from us so that there exists a moral order between both. As argued in indigenous and various Asian, particularly Buddhist, societies (Sperber 1995, Sandell 1995) this relation implies that there would be a reaction of nature to human failings. Similar notions surface even in the secular West, in the argument that nature were striking back against us with extreme climatic events and the like. There are also common ideas of a reaction of nature to extraordinary persons and/or circumstances (such as the birth and enlightenment of the Buddha or the crucifixion of Jesus Christ; maybe the reaction of the Matrix to Neo's special powers in the movie falls into such a category as well).

Social as well as ethical considerations are, moreover, forming a main backdrop to sustainability's orientations on intra- and intergenerational equity, i.e. present social justice and an upholding of possibilities of future generations for survival and for deciding on which ways of life to follow.

Potentially, thinking towards the future, an orientation on such issues could also provide a positive contribution. Certainly, with regard to moral/ethical considerations of the global human situation, and with related effects on security that could easily come to affect even the better-off if life chances for too many others are diminished (and may already be doing so, cp. Ch. 14), this would likely be the case. Another contribution to a good life could come from a shared understanding of our planetary citizenship and common humanity (cf. Singer 2004, for example), and globally common (even if locally differently expressed) "ecological identity," i.e. "how people perceive themselves in reference to nature, as living and breathing beings connected to the rhythms of the earth, the biogeochemical cycles, the grand and complex diversity of ecological systems" (Thomashow 1995: xiii).

Moralistic valuations, finally, are related to social needs through the "strong feelings of affinity" that form a part of both (Kellert 1993: 53). These relationships reach from the humanistic valuation of nature into the moral/ethical consideration which is a major focus in the explicitly non-utilitarian discussion of what is coming to be called "spiritual ecology" (Sponsel 2001).

Chapter 18

Spirituality

Religion would have been related with Maslow's suggested need for self-actualization and the like, which he dropped in later revisions (cp. Anderson 1996). So, it is not strictly fitting for the framework of human needs to include religion. As a distinctly human form of "knowledge," which has only recently been coming to be strongly implicated in environmentalism, however, systems of belief are well worth mentioning. Moreover, they are among the strongest motivations for a considerable number of people, and all the more interesting for providing a perspective that is yet more removed from the utilitarian, and even arguing with and for the eminent meaning of the non-utilitarian to human happiness.

In their roles as soteriology, in legitimating social relations, etc., religions do not have the strongest of relations to the environment. There is, however, a fundamental similarity between the questions of religion and human ecology/environmentalism that pertain to the workings of the world and the human position and role in it (Kinsley 1995: xv f.). It can even be argued, from the perspective that religious worldviews are among the most basic such views that shape human action, that their role in human ecology is fundamental. So, their understanding is a necessary, and so far largely lacking, ingredient to the understanding and potential transformation of human society's interaction with the environment (cp. Sullivan 1997).

A more direct relation can be argued to exist: nature has spiritual value as a setting and object for contemplation and spiritual growth quite beyond its value as a relaxing and/or arousing environment. It can even be considered (not only rightly, the naturalistic fallacy of interpreting that what is natural were right and good is always close by) to be a model and teacher to religious experience, whether as (book of) creation (Abrahamic), paragon of impermance (Buddhist), model of unity of heaven, earth, and humans (Confucian), etc.

The diversity of perspectives at the interface of religion and ecology, a representation of which can also be found in "Cultural and Spiritual Values of Biodiversity" edited by Posey (1999), has been discussed in most detail as well as in the widest latitude in the series "Religions of the World and Ecology" published by Harvard University Press: Grim (2001) on indigenous traditions, Chapple and Tucker (2000) on Hinduism, Chapple (2002) on Jainism, Tucker and Williams (1997) on Buddhism, Tucker and Berthrong (1998) on Confucianism, Girardot, Miller and Liu (2001) on Daoism, Tirosh-Samuelson (2002) on Judaism, Hessel and Ruether (2000) on Christianity, Foltz, Denny and Baharuddin (2003) on Islam, Bernard (forthcoming) on Shinto, and Tucker and Grim (forthcoming) on "Cosmology and Ecology."

Notwithstanding the philosophical and practical conundrum that spiritual considerations can cause, particularly when the naturalistic fallacy holds its sway or when ideologies cloud the perception of what is going on empirically, moral and ethical aspects do form important considerations best addressed in the present context, again.

The moralistic valuation of nature, besides its expression in affinity and emotional attachment to (features of) nature shared with the humanistic valuation, is also shown in "the conviction of a fundamental spiritual meaning, order, and harmony in nature ... sentiments of ethical and spiritual connectedness" (Kellert 1993: 53). One manifestation of these valuations, as mentioned in the above chapter, lies in the more anthropocentric social, ethical considerations of the linkage between sustainability as upholding of the ecosystemic foundations for human life, and concerns, even needs, for a move towards social, intra- and even intergenerational justice. Yet less anthropocentric, more spiritually oriented considerations are to be found in the discourse on the intrinsic value of nature.

Such spiritual ecologies are by and large distinctly modern developments arising both out of an age-old concern over the proper understanding/teaching of the relationship between humans, nature and the supernatural, and the mainly recent environmental crisis. This "complex and diverse arena of spiritual, emotional, intellectual, and practical activities at the interface of religions and the environment" (Sponsel 2001: 181) has been developing rather reluctantly at first, mainly during the last decade, but recently gained prominence. After all, religious thought, whether ancient, contemporary, or a re-interpretation of the ancient in light of modernity, still holds considerable interest, at the very least as a pool of human ideas. Not quite as rarely as the modern notion of an advancing secularism would have believed, religion has continued to hold or even reasserted its special place among humanity – if not always to the better, unfortunately. In pointing out that there may be more to human life than just the banal, secular – possibly fulfilling what may be a truly human need for ideas of transcendence and meaning – religions are among the cultural features at large that are most subversive of the ideology of consumptive materialism: They actually dare to hold on to the view that human beings are not islands unto themselves but intimately bound up with others, in and with the world; that happiness in life does not lie with how many products of what kind you can and do afford, and that there are other, deeper truths beyond the turmoil of daily life.

Chapter 19

Nature, Knowledge, and Creativity

Receiving relevant information of the environment and managing it appropriately in order to survive and reproduce is a main need for any species, of course. Learning, social transmission of knowledge, does also play a role for many social species' basic skill set, to the point of different chimpanzee groups developing different "cultural traditions." In homo sapiens, the natural need for information management is so much more relevant, however, that reasoning is the feature for which our biological designation was created; "sapiens" meaning reasonable, understanding, intelligent.

The socially transmitted knowledge is one of the most prominent factors of human life – the diversity of cultures builds up on it, creating a second 'evolutionary' system (biology being the first, of course) without which even basic survival would not be possible, but which includes many factors that are not directly related to survival, nor necessarily to reproductive success, at all times (Durham 1991 delivers an excellent discussion of the interrelations between inherent biological "primary values" such as survival and reproduction, "secondary values" of culture, and cultural features, as well as the positive, negative, or neutral relationships between genes and culture that occur in the relation).

Ultimately, the increasing role of social dynamics and technology has made the cultural and technological environment so important that it is considered the more relevant basis for life, especially recently; witness the view that the economy were not based in/on nature any longer, taken to the extreme in the economic "virtualism" that would want to make the world conform to the virtual reality constructed by the idealization of the free market (Carrier & Miller 1998), and the less high-flight observation that, naturally, a majority of people nowadays derive their livelihood not in direct interaction with the natural environment.

So, it would come as no surprise that the knowledge the acquisition of which is culturally supported to the largest extent, and extant among most people, is not pervasively concerned about the natural environment (cp. Wolff & Medin 2001). – In fact, it can even be so minimal that having or not having goldfish as pets has measurable effects on children's biological reasoning (Hatano & Inagaki 1987; Inagaki 1990; referenced in Wolff & Medin 2001).

In contrast, among (indigenous) peoples with subsistence economies based on livelihood derived from more direct interaction with nature, "naturalistic" and "ecologistic-scientific" valuations – knowledge, curiosity, and understanding of nature – would not have been just values, but necessities of making a living. Knowledge of the surrounding environment and of how it can best be utilized is so well developed among many, if not most, such peoples that it shows considerable overlap with, and in many cases and regards even better/wider understanding than, modern science – cp. the discussion in ethnobiology and ethnoecology, for example. As a matter of course, it also includes spiritual, mythological ideas which may not be scientifically (but culturally) valid; anyway, since the ethnocentrism inherent in a valuation of this "other" knowledge in terms of its closeness to science is apparent, this is not the point. The importance lies with cultural support on the one hand, and the ultimate importance of (knowledge of) nature to many peoples and eventually all of us, on the other hand.

Even in contexts less immediately related to survival, nature has been having a special position in and for human thought. As Lévi-Strauss had put it, animals for example are not only good to eat, but also "good to think." The "symbolic" value of nature may have been essential in human development, as it is providing symbols for thought such as the (actual and/or attributed) properties of animals (cp. Lawrence 1993), and natural metaphors in general, seeing as natural processes are "a rich resource for the construction of meaning and expression of feelings" (Bruun & Kalland 1995: 4). It would be tempting to say that the ecologistic-scientific valuation of nature, expressed in systematic study – i.e. nowadays in science – would at least have taken a turn to the better. Understanding of global cycles, for example, has certainly improved (or even only begun to be possible rather recently), but the interest in and meaning to daily lives has decreased with technological advances.

In spite of the more limited contemporary interaction, nature continues to serve such a function, at least in metaphors and figures of speech (but as noted in Chapter 17, oftentimes in ways which are rather detrimental to the relationship between cultural support and actual protection). In spite of what appears to be a rather (pre- and) historic relationship when artificial environments are more pervasive nowadays (and even natural areas not only shaped by non-human nature), there are indications that greener as well as wilder environments still hold great and special relevance to human cognition and knowledge, both to scientific/theoretical and to applied concerns.

In children's physical and mental development, for example, contact with nature continues to have great potential for contributing positively, e.g. to psychomotor skills, social interaction, concentration (cf. Kahn & Kellert 2002). In contributing to adult, modern people's cognitive functioning, as a resource in and with which to develop physical and even cognitive skills, in promoting technological and creative development(s), as well as for scientific understanding, there is a number of roles that nature still is (capable of) playing, too.

Cognitive Functioning and Nature

Corporeality once again is an aspect worth of pointing out explicitly, particularly as the body is not often considered as a part of, and as, nature. The relation in the current context comes to the fore particularly since rest/arousal impact on the body's general constitution, influencing cognition via concentration, for example. The need for sleep (to connect with the example of rest and arousal again; cp. Ch. 16), too, interacts with this issue – it is a natural influence on learning (memory, cp. Maquet 2001), and even a potential contribution to coming up with new solutions (cp. Wagner et al. 2004). Interaction between cognition, the world, and the person also has to be realized to take place through the body. – Even cognitive and emotional processes change, and in some cases change with, the body; the brain is but one part of this complex (Kutas & Federmeier 1998).

Thus exercise, for example, has a salutary – preservative and restorative – effect on some areas of cognitive functioning across the life-span (Woo & Sharps 2003; Bashore & Goddard 1993). Anecdotally, an even more direct relationship is suggested since most insights appear to have been gained not in the actual study, but pondering the questions (or not even assuming to be doing so anymore) while perambulating, day-dreaming, or the like, oftentimes in pleasant, green surroundings. Certainly, these considerations relate active lifestyles that also contribute to bodily health with sustainable lifestyles, reducing consumption while gaining a non-material "profit" from nature.

Cognition is also supported by "green" environments (surroundings), as relaxation and arousal (which in turn contribute to cognitive functioning) are. For example, children's attention capacities were significantly improved by "greener" surroundings to their home (Wells 2000); "nearby nature" in view from homes is a factor contributing to girl's self-discipline (concentration, impulse inhibition, and delay of gratification), though not to that of boy, for whom nature further away is likely to be more important, and of positive effect as well (Faber Taylor et al. 2002). The relation in children is particularly striking with regard to ADD (attention deficit disorder): what is a great problem in school and home settings appears to be quite adaptive, exploratory behavior in natural settings, and nature experience can, as just described, contribute to attention even in children with ADD (cp. Faber Taylor et al. 2001).

The relationship holds true not only for children, but for adults as well – rest/arousal are enhanced by natural features/environments, attention is restored by it (e.g. Tennessen & Cimprich 1995), in effect better work is done in better, including more natural, green, environments (cp. Kaplan, R. 1993; Hawken, Lovins & Lovins 1999).

Science, TEK, and Sustainability

The naturalistic and ecologistic-scientific valuations of nature, which are less important to the majority of people living nowadays except for those who are amateur naturalists or birdwatchers (or of course scientists), have been finding their main – also utilitarian – expression in traditional ecological knowledge (TEK) and modern natural (mainly biological and ecological) science. The promotion of scientific understanding clearly needs diversity, actually both biological for the (natural) sciences and cultural, e.g. linguistic, for the humanities; and eventually even understanding and promotion of "our common humanity" and of our relation as beings in the (same) world will require diversity (cp. Harmon 2001).

Not only does natural diversity contribute to this form of knowledge which is rather detached from the interests of many people, however. The diversity, richness in information, and occurrence of quirky, unexpected elements is an input to human understanding and creativity in general, as well. It goes beyond and influences the variety in human creations. Terentius' saying "homo sum, nihil humani a me alienum puto" (I am a man, therefore nothing human is alien to me; incidentally, he was a Roman writer of comedies) thus exhibits the downside that human imagination without inspiration by natural diversity and quirks is rather more limited than it is with them (witness the limited diversity of TV shows?).

The above-mentioned content of culturally but not scientifically valid knowledge, e.g. mythologies, religious metaphors, stories, sentiments, about nature here comes to the fore: Typically they exhibit a deep, but oftentimes hidden, relation to the natural world. Rather than being mere creatures of imagination, there are influences, impulses given by nature – whether it were the idea of plants that could protect from sprites (which is particularly interesting when it exhibits connections with medicinal knowledge or parallels in different countries that cannot be attributed to contact), mythological monsters, or possibly H.R. Giger's Alien (cp. Wickler & Seibt 1998, though not for this last example). The interest in cryptozoology probably attests to this relation of creativity, interest and nature as well; and a biologist trying to explain his/her fascination would likely concur that a looking glass and a patch of grassland is enough to find a few odd creatures.

Wilson (2002: 146; not surprisingly) supports the idea of the special standing of nature, stating that: "To the multiple valorizations of wild environments can be added mystery. Without mystery life shrinks. The completely known is a numbing void to all active minds. Even a laboratory rat seeks the adventure of the maze."

It is not necessary to go into such "esoteric" connections, yet: The natural world represents a treasure chest of ideas for technological development(s) that has only recently started to be tapped as well, e.g. in bionics and as an influence on ideas of industrial ecology, ecological design, and the like. Artifacts in general can just as cultural ideas be argued to stand in a relation to life that is deeper than commonly considered, not just because of the impact of consumerism but for the inspirations it provides (cp. Gadgil 1993).

Ultimately, even considering ideas for long-term space exploration and colonization, at least as much ecological as technological understanding will be required: Physical aspects such as the most efficient systems of propulsion, geochemical aspects such as those required for the transformation of an atmosphere in "terraforming," or biological-genetic effects of zero gravity and cosmic rays are being discussed, but the salient (probably technological-) ecological requirements for providing food, clean water, air, etc. in a small closed system, let alone for terraforming, are finding much less prominence. The one experiment concerned with such latter aspects, Biosphere 2 at best (may have) contributed to the understanding that the Biosphere 1, i.e. Earth, is a wondrously functioning home that – rather than dream only of where to go next – we should take better care of.

In summary, nature and knowledge interrelate in several ways, respectively as a source of ideas and fount of knowledge, a setting contributing to cognition, in the one direction, and as the pool of traditional as well as modern ecological knowledge useful, the other way round, as a key towards managing, living with and off ecosystems sustainably; even (and particularly) if, under the conditions of higher population numbers and with the wish/need for technology, sustainability will require a lot of fresh thinking.

Knowledge, in a context stronger relating it to relevance to the individual, is a precondition for even coming to the idea that satisfaction could be derived from certain lifestyles, including those giving more importance to active living and contact with nature. For example, that ways of life more oriented towards sustainability, as in the contexts of nutrition and health, rest/arousal, are available and likely contributing to well-being has to be at least suspected to be given a try, and then some ideas for what to do need to be available. Even where the enjoyment of foods that the diversity of specific cultivars can bring is concerned, for example, one needs to know that there is a range quite outside the apparent variety readily available in supermarkets (and where to find it). Knowledge of specific uses more appropriate to certain varieties than to others, etc., helps in the enjoyment, too, as may ideas of the relation between the way of life represented by this behavior and social justice or sustainability contribute to the sense it makes for the individual.

The relationship between nature, cognitive functioning, and knowledge, then, can be mutually supportive once initial inertia has been overcome: When it contributes to concentration, for example, this in turn improves the possibility for additional discoveries, experiences, and knowledge, which enhances awareness and the success of the experience yet further.

So is cognition of the greatest importance for individual identity, which is a major area of possibilities for individual decisions, and a major problem because of the sheer range of possibilities: In a world where identities are less inherited by birth than having to be invented, a realization of oneself and one's place in the immediate environment is a good starting point from which exploration by scientific knowledge and imagination can work, e.g. to come to understand one's "ecological identity" (Thomashow 1995, 2001), and decide to shape one's identity, self-perception, way of life, and role in the world.

In this context, the diversity of cultural knowledge, not only of TEK for sustainably living with ecosystems, but also on what makes life worth living, is an asset worth pointing out explicitly: People are (although similar in basic needs, for example) different in terms of what is most important to them. Most of the time, environmentalist discourse quarrels over the "silver bullet"-solution, but it should finally be understood that this diversity could be a positive factor, if only mediated by being connected with the orientation on the common good of sustainability. Of course, although the respective scientific studies aim for general patterns, such individual differences exert great influence. Thus, diversity of approaches is also (beside its role towards resilience, cp. Ch. 7) an important aspect for this reason.

The challenge, which becomes especially salient in this context of the precondition of knowledge (not only as information but also as experience, meaningful translation of information into usable guidance), is that such education does require some initial impulse, but its competition with entertainment and consumerist promises in an open market is a competition between "easy" and "hard," and "for those not yet disciplined in the rites of learning, 'freedom' will always mean easy" (Barber 1995: 117).

The problem is apparent in the quote by Wilson (above), since the rat may seek the adventure of the maze, but many people caught in the "rat race" of modern life (supposedly) have but little possibility, and (more importantly) little impulse to seek other arousal but that most strongly, culturally, supported – such as consumption of media and products. In contrast, the restorative potential of nature, for example, is very likely to be of actual effect in just about any person, but "when choosing a restorative activity..., especially when choosing for oneself, nature activities are underappreciated relative to entertainment activities" (Herzog et al. 2002).

As Barber further, in an ultimately seething note, puts it, "to grow into our mature better selves, we need the help of our nascent better selves, which is what common standards, authoritative education, and a sense of the public good can offer. Consumption takes us as it finds us, the more impulsive and greedy, the better" (1995: 117).

Active living – both as personal decision on one's way of life based on what is individually relevant rather than what is supposedly necessary, the fashion, and the like, and as active doing rather than passively consuming – would, in general, however, likely be correlated with increasing well-being also because of its relation to the need for self-efficacy (which is now to be addressed).

Chapter 20

Sustainability and Self-Efficacy

It is not often recognized as such, but there is an actual need for feeling to have some control over one's life. A consumerist economy functions (relatively) well marketing substitutes (and/or diversions) particularly as long as there are enough people with jobs (which satisfy that feeling as well, which is one of the main reasons, besides the obvious loss of income, why unemployment is psychologically challenging). It works because it builds on basic impulses (cp. Barber's comment quoted in the above chapter), but has been going so far as to lead critics to argue that people's capacity for coping with the ups and downs of life is systematically being undermined – even to the point of medicinal ideas being used for social control (cp. Fitzpatrick 2000, as well as to sell "lifestyle drugs") rather than with pertinent (support for) practical or psychological action to really deal with the situation.

The challenges thus go deep, as self-efficacy, the effort it takes, and its connection with sustainability are but little supported, but everyone by the very fact of being alive is also of influence on life and sustainability. This understanding might come a long way in and of itself, but is commonly denigrated – unsurprisingly, when the alternatives are presented as bleak existence or are not presented at all – in favor of various forms of denial (cp. Opotow & Weiss 2000). As such a major arena for denial or a recovery of the sense of control by taking action, the practical and/or psychological dynamics related to this need for competence directly bear on the issue of futility/agency – hence leading full circle to the introductory comments.

Any more active approach to one's life, however, is likely to contribute to the sense of personal control/influence over one's life. There are, of course, some people who will explicitly recover their (sense of) agency in the world by activism or action, e.g. in environmental(ist) or conservationist contexts. Primack (1998: 5) sounds an interesting note in this regard: "Some people feel discouraged by the avalanche of species extinctions..., but it is possible to feel challenged, instead... People may someday look back on the ... time when a relative handful of determined people saved numerous species and some entire biological communities." – Similarly, the "translation" of sustainability into locally, not only environmentally but also culturally (and individually) fitting, meaningful action still requires a lot of effort – and at the same time offers itself excellently to a recovery of the sense of agency and social/cultural identity (even while upholding a sense of inclusion in global humanity or culture).

Nature, whether it be wilder, far away, or more controlled and close-by, once again offers some of the best, most direct possibilities for recovering an individual sense of self-efficacy. Again, such processes are already in full swing during children's development, presupposing that natural areas are available, e.g. in exploratory behavior and, even more so, in the creation of refuges, "imaginary play, a behavior with high social and cognitive payoffs ... facilitated by natural refuges and natural materials" (Heerwagen & Orians 2002).

Also following the pattern seen in preceding chapters, adults accrue similar benefits such as self-assurance and confidence from outdoor/wilderness programs (as reviewed in Kaplan & Kaplan 1989; with enduring effects, cp. R. Kaplan 1974, 1977), for example, among other things as a likely result of the immediate feedback between action, skill, and successful outcomes.

The environmental-activist, as well as educated consumer, emancipated citizen, etc. activities also would contribute to the feeling of necessary control over (or less strictly put, influence on) one's life, as the related "skills and abilities useful in taking care of the planet" also serve to provide (both feelings of and actual) competence (De Young 2000); the social and psychological factors effective in the context of sustainability still needed to be brought to bear more effectively, but could "counter the pervasive malaise of happiness," and "by providing opportunities for understanding, exploration, and participation, effective group problem solving [for example] can lead to new multiply desirable choices," rather than further – demotivatingly – emphasize "known necessary sacrifices" (Kaplan 2000). Furthermore, it is helpful to note that an orientation on life experiences in general has been shown to contribute more to a good life than the materialistic approach of consumption (cp. Van Boven & Gilovich 2003).

Ways of more actively shaping one's life would also exhibit political relations, as everyone is, as the adage goes, either a part of the problem or a part of the solution. Here, however, this also interacts with a notion of the pursuit of happiness, true freedom, true diversity meaning not only the "freedom" to choose product A or product B, but distinct ways of life, individual and collective values, well-being and the common good. It does, however, require an understanding that, as not everything is possible for physical reasons or because of laws, so sustainability will require that ecological concerns influence what can and should be done. The trouble is that in aerodynamics, for example, the patterns it is necessary to work with (i.e. the natural laws of physics) are obvious, whereas the ecological patterns that it will be necessary to work with towards sustainability are rather systemic, i.e. less obvious, oftentimes more forgiving for longer times, but ultimately none the less real.

On the other hand, the struggle towards cultures of sustainability can at least be understood to be a "moral equivalent of war," as "a superordinate goal that all nations and peoples can share" (Oskamp 2000), in which every person has a role to play – and in contrast to war, not become a victim but see positive effects, as this work hopes to have started to show. Ultimately, although we tend to forget it, "people gain .. intrinsic satisfaction .. from being competent at doing things that have a positive effect in a larger context and that matter in the long run" (De Young 2000: 522), which is a great cause for hope.

Conclusion

Change has always been a constant of life, not just in this age. Changes that are coming or even under way now appear to include a newly rising importance of resources, especially of the living kind and of fossil fuels, environmental changes, among them the lack of sinks which is a far more pressing problem with regards to most resources than their limits, the state of ecological services affected by global change(s), but also economic and communicative globalization which includes what may amount to a long called-for redistribution of income between countries, and finally the new population transitions.

Interestingly, all changes that fit in with the ideology of growth, e.g. fossil fuel/energy consumption in spite of global change and limits, a new return to population growth in industrialized countries which currently have little population growth (suggested as solution to their population transition), off-shoring of jobs, are near unequivocally portrayed as ultimately good for the economy, and therefore supposedly good for everyone. All the suggestions of necessary changes toward sustainability, in contrast, are portrayed as detrimental. Moreover, there are still strong tendencies of resistance to the latter kind of change, even of outright denial of its necessity; and industrial-economic, consumerist "development," that is economic growth, is being presented (as well as taken) as the only way towards happiness. This simplistic ideology is easy to understand and appears justified by the "North's" richness, but it has come at a cost and is not a model that the whole world could emulate.

Ultimately, however, creating ways of living that constitute an integrated and even ecologically restorative economy, even if (in practice) ignored so far, by and with an orientation shifting from mere short-term profit derived from natural capital and by the psychology of the stock market to sustainably living off the interest will become necessary if a positive development is to be achieved. Some consumption, definitely as necessary to satisfy basic needs, is required and has to be (made) possible, but apart from that base, happiness does not come from increasing consumption/economic growth but is a highly individual matter that has at least as much to do with intrinsic motivation, experiences, and ways of life as with material goods (cp. Eckersley 2000, Bowling & Windsor 2001, Kasser & Kanner 2004).

Searching for the single, perfect solution – as essentialist tendencies in environmentalism tend to lead to – in any one of the relationships described above, this work would have failed in a similar way to just how environmentalism is in some trouble (as argued by Shellenberger & Nordhaus 2005). Considering that "saving the

Earth" is not enough of a motivation for a transformation to sustainability, as well as the diversity of human motivations themselves, however, this is hardly surprising. Deliberately, this work tries not to present a (necessarily, then) ideological "one for all"-solution. It is not even necessary, or likely, that any and all of the relations described hold true for everyone; yet, as findings of science that strive for general insight, it is likely that they would hold true for most people, so should be worth the try. This is one of the reasons why it was tried to focus more on approaches that would uphold a wider range of possibilities than "dominant" approaches would (the term dominant usually being reserved to add economic, but applicable to environmentalism as a special interest issue as well). The highly individualistic character of what constitutes a "good" life further purports to the need for an orientation not on happiness directly, but on possibilities – on livelihood and life chances.

The picture that is emerging nevertheless is supportive of the initial hypothesis, the perspective of positive ecology that ways of living that are contributive to both ecological patterns (e.g. biodiversity) and processes (e.g. ecosystem services) and human survival and well-being could be found by integrating the different (but actually interrelated) concerns of ecological sustainability and an anthropology and psychology of human needs and the "good life," informed by and oriented on the relevance and interconnection of various, universal relationships, values, and especially human needs and appropriate means to their satisfaction.

In the likely diversity that good lives would take, related to the diversity that is inherent in what needs and values are most important to different people, for example, there is a positive correlation with the necessity for diversity as an evolutionary precondition to ongoing development. This presents the confluence that is repeatedly the case: Any one factor taken per se can probably not be satisfied in a perfectly "positive ecological" way. Yet by way of their interrelation, this orientation would offer positive synergies where currently dominant ways of satisfying needs – and commonly marketing proxies for needs while inventing wants that are portrayed as essential necessities – impose the costs of externalities on society and the ecosphere at large.

Sustainability and Happiness

Drawing the different strands that run through this work together, three ways of conceptualizing nature – following Kaplan and Kaplan 1989, as an amenity (of wild biodiverse nature), as a necessity (of ecosystem services), and as an essential bond (considering the "practical" and psychological entirety of human integratedness in the world) – can be described and considered to shape three different scenarios for the future. These particular scenarios are taken from Murray & Cook (2002), where they are used only to illustrate China's possible futures, but in their general thrust they appear equally applicable to the world at large, and all the more interesting considering that this work has been hailed as one of the few balanced analyses (apparently, meaning ones that allow for the possibility of a way out of the trouble?). In their alternatives, these scenarios show the necessity and/or potential of a positive ecology yet better:

1.) Nature "viewed as an amenity, [which] may be readily replaced by some greater technological achievement" (Kaplan and Kaplan 1989: 203), relates to the "gloomy, ... more realistic" (Murray & Cook 2002: 213) scenario of business-as-usual. That is, concern over environmental issues is not seen as necessary enough for more than a little cleanup to occur. Thus, in total there will be too little (and too late) to be of much more than cosmetic effect. An "anthropic nature" consisting near exclusively of urban and agricultural lands yet proves to be enough for human life support, however, and problems such as alternative sources of energy either did not materialize or were solved. Many or most of the basic relations (as described in Part III) need technological replacement (which is costly but here considered practicable), and the eco-cultural relations (as described in Part IV) only remain in force in fragmentary form. Life goes on – rather like it was described in the cyberpunk literature, however, i.e. rather bleak (unless if one thought that none of the relations described here held true, – but a change towards the second scenario could still occur).

2.) Nature as a necessity can be related to Murray and Cook's "doomsday scenario" (2002: 216ff.). The initial processes would go basically as in the above scenario, with little concern over the environment. But as the impact of humanity on the ecosphere and/or biodiversity in detrimental ways increased, the assumption fundamental to this scenario is that it would produce a catastrophic shift in ecosystem/ecosphere states and, hence, of living conditions. It would not even take as extreme an event as the breakdown of Northern Atlantic thermohaline circulation, then, to achieve similar negative effects of all-out strife for natural resources, food and water, etc. To put it simply and mildly: Life chances, understandably, would be considerably lower.

3.) Viewed (also) "as an essential bond between humans and other living things, the natural environment has no substitutes" (Kaplan and Kaplan 1989: 203). The views as necessity, amenity, and essential bond formed the backdrop to this work in its different parts, arguing that all of them were of effect and importance. Therefore, they would all need to be heeded. Were this achieved, and actually "translated" into practice (which is a major problem both of academic and of activist approaches to sustainability, admittedly) as strongly as necessary, basically building up "cultures of sustainability" in the wide sense of the term on their true, primarily ecological and secondarily human foundations, a "rosy" scenario (cf. Murray & Cook 2002: 210ff.) could materialize: Life chances are enhanced as there are more possibilities of how to live, (in spite of as well as because of our) making a living in ways that are not detrimental to ecological sustainability but working with and as a part of nature.

In light of the relations described, it is hoped that the perspective of "positive ecology" has become clear: Rather than an environmental orientation implying austerity and bleakness, this could easily be a result of continuing as if no change were needed. Taking appropriate, empirically-validated perspectives on human needs and their fulfillment, together with their relation to nature, into account, the space of possibilities would actually be expanded by an orientation on sustainability as the co-evolution of humanity with (and as the part having to be more considered) and in nature. Thus, a contribution towards progress, defined in a wider sense than just as economic growth, e.g. as promoting possibilities for development as well as towards a good life, chances for personally relevant ways of life, new experiences, the pleasures of true diversity, and so much more, would be gained by it.

The challenge of the required changes – "doing things as never before, and doing without" (Smil 1994: 108), without thoughtless consumption at least, – does present itself as obstacle of greatest difficulty. *Ad hoc* reaction rather than a proactive, balanced perspective unfortunately holds its sway. Yet, human history is indeed one of overcoming obstacles and changing perspectives – albeit also one of failure of individual communities and survival as a species, which may be a "more pertinent [perspective] in future research," as argued by Emilio Moran (2000: 338). Hopefully, we will finally turn towards the proactive, more inclusive, and also more long-term, striving to realize our potential of being the sapient species.

Bibliography

Ahuvia, Aaron C. "Individualism/Collectivism and Cultures of Happiness. A theoretical conjecture on the relationship between consumption, culture and subjective well-being at the national level." *Journal of Happiness Studies* 3 (2002): 23–36

Alexander, Susan E., Stephen H. Schneider, and Kalen Lagerquist. 1997. "The Interaction of Climate and Life." In: Daily, Gretchen C. (ed.) *Nature's Services. Societal Dependence on Natural Ecosystems*. Washington D.C. and Covelo, CA.: Island Press. Pp. 71–92

Altieri, M.A., S. Hecht, and R. Norgaard. 1987. *Agroecology. The Scientific Basis of Alternative Agriculture*. Boulder: Westview Press

Altieri, Miguel A. 1999. "The ecological role of biodiversity in agroecosystems." *Agriculture, Ecosystems and Environment* 74 (1999): 19–31

Anderson, Eugene N. 1996. *Ecologies of the Heart. Emotion, Belief, and the Environment*. Oxford, New York: Oxford University Press

Anderson, Kat and Gary Paul Nabhan. 1991. "Gardeners in Eden." *Wilderness* 35 (194): 27–30

Anson, R. Michael, Zhihong Guo, Rafael de Cabo, Titilola Iyun, Michelle Rios, Adrienne Hagepanos, Donald K. Ingram, Mark A. Lane, and Mark P. Mattson. "Intermittent fasting dissociates beneficial effects of dietary restriction on glucose metabolism and neuronal resistance to injury from calorie intake." *PNAS*, Vol. 100, No. 10 (13 May 2003): 6216–6220

Arrow, K., B. Bolin, R. Costanza, P. Dasgupta, C. Folke, C.S. Holling, B. Jansson, S. Levin, K. Maler, C. Perrings, and D. Pimentel. 1995. "Economic Growth, Carrying Capacity, and the Environment." *Science* Vol. 268 (28): 520–521

Atran, Scott. 1993. "Itza Maya Tropical Agro-Forestry." *Current Anthropology*, Vol. 34, No. 5 (1993): 633–700

Balmford, Andrew; Aaron Bruner, Philip Cooper, Robert Costanza, Stephen Farber, Rhys E. Green, Martin Jenkins, Paul Jefferiss, Valma Jessamy, Joah Madden, Kat Munro, Norman Myers, Shahid Naeem, Jouni Paavola, Matthew Rayment, Sergio Rosendo, Joan Roughgarden, Kate Trumper and R. Kerry Turner. 2002. "Economic Reasons for Conserving Wild Nature." *Science* Vol. 297 (9 August 2002): 950–953

Bandura, Albert. 1982. "Self-Efficacy Mechanism in Human Agency." *American Psychologist* Vol. 37, No. 2: 122–147

Barber, Benjamin R. 1995. *Jihad vs. McWorld. How Globalism and Tribalism are Reshaping the World*. New York: Times Books

Bashore, T. R. and P. H. Goddard. 1993. "Preservative and restorative effects of aerobic fitness on the age-related slowing of mental processing speed." In: Cerella, J., J. M. Rybash, W. Hoyer, & M. L. Commons (eds). *Adult Information Processing: Limits on Lss.* San Diego: Academic Press. Pp. 205–228

Benson, John F., and Maggie H. Roe (eds). 2000. *Landscape and Sustainability.* London and New York: Spon Press (Taylor & Francis Group)

Benyus, Janine M. 1998 (Reissue 2002). *Biomimicry. Innovation Inspired by Nature.* New York: Perennial (HarperCollins Publishers)

Bernard, Rosemarie. forthcoming. *Shinto and Ecology.* Cambridge, MA: Harvard University Press

Bernard, Ted, Jora M. Young, and Wes Jackson. 1996. *The Ecology of Hope: Communities Collaborate for Sustainability.* Gabriola Island, B.C.: New Society Publishers

Berton, Valerie (ed.) 2001. *The New American Farmer. Profiles of Agricultural Innovation.* USDA – Sustainable Agriculture Research and Education program (USDA-SARE)

Bogin, Barry. 1991. "The evolution of human nutrition." In: Romanucci-Ross, L., D.E. Moerman, and L. R. Tancredi (eds). *The Anthropology of Medicine. From Culture to Method (Second Edition).* New York: Bergin and Garvey. Pp. 158–195

Botkin, Daniel B. 1990. *Discordant Harmonies. A New Ecology for the Twenty-First Century.* Oxford, New York: Oxford University Press

Botkin, Daniel et al. 2000. *Forces of Change. A New View of Nature* (Smithsonian Institution/ National Museum of Natural History). Washington, D.C.: National Geographic Society

Boucher, Douglas H. (ed.) 1999. *The Paradox of Plenty. Hunger in a Bountiful World.* Oakland, CA: Food First Books

Bowling, Ann and Joy Windsor. 2001. "Towards the Good Life: A Population Survey of Dimensions of Quality of Life." *Journal of Happiness Studies,* 2: 55–81

Braun-Fahrlander, C., Riedler J, Herz U, Eder W, Waser M, Grize L, Maisch S, Carr D, Gerlach F, Bufe A, Lauener RP, Schierl R, Renz H, Nowak D, von Mutius E; Allergy and Endotoxin Study Team. 2002. "Environmental exposure to endotoxin and its relation to asthma in school-age children." *New England Journal of Medicine,* Vol. 347, No. 12 (19 September 2002): 869–877

Brown, Lester R. 2001. *Eco-Economy. Building an Economy for the Earth.* New York, NY: W. W. Norton & Co.

Brownell, Kelly D. and Katherine Battle Horgen. 2004. *Food Fight. The Inside Story of the Food Industry, America's Obesity Crisis, and What We Can Do About It.* Chicago and others: Contemporary Books (The McGraw-Hill Companies, Inc.)

Bruun, Ole and Arne Kalland (eds). 1995. *Asian Perceptions of Nature. A Critical Approach.* Richmond: Curzon Press.

Bryld, Erik. 2003. "Potentials, problems, and policy implications for urban agriculture in developing countries." *Agriculture and Human Values* 20 (2003): 79–86

Buchmann, Stephen L. and Gary Paul Nabhan. 1997. *The Forgotten Pollinators.* Shearwater Books

Cairns, J. 2000. "Sustainability and the Future of Humankind: Two Competing Theories of Infinite Substitutability." *Politics and the Life Sciences*, Vol. 19 (1 March 2000), No. 1: 27–32

Campbell, Neil A. and Jane B. Reece. 2001. *Biology (6th edition)*. Pearson Higher Education

Cantril, H., 1966. *The Pattern of Human Concerns*. New Brunswick, NJ: Rutgers University Press

Carrier, James G. and D. Miller (eds). 1998. *Virtualism: A New Political Economy*. Oxford: Berg

Cary, J., 1998: Institutional innovation in natural resource management. In: Australia. The triumph of creativity over adversity. In: Abstracts of the Conference "Knowledge Generation and Transfer: Implications for Agriculture in the 21st Century." Berkeley: University of California-Berkeley

Cervinka R. 1993. "Nightshift dose and stress at work." *Ergonomics*, Vol. 36 No. 1–3 (1993): 155–60

Chapman, L. T. Johns, and R.L.A. Mahunnah. 1997. "Saponin-like in vitro characteristics of extracts from selected non-nutrient wild plant food additives used by Maasai in meat and milk based soups." *Ecology of Food and Nutrition* 36: 1–22

Chapple, Christopher Key (ed.) 2002. *Jainism and Ecology. Nonviolence in the Web of Life*. Cambridge, MA: Harvard University Press

Chapple, Christopher Key and Mary Evelyn Tucker (eds). 2000. *Hinduism and Ecology. The Intersection of Earth, Sky, and Water*. Cambridge, MA: Harvard University Press

Chatel, Thomas and Gernot Steinweg. 2002. *Wasserkrieg in Spanien / Espagne: la guerre de l'eau*. ARTE Reportage (TV documentary), transmission on 03–28–2002, production by ARTE GEIE, Picture Pan, Germany 2002. Online information on www.arte-tv.com

Clark, Colin W. 1973. "Profit maximization and the extinction of animal species." *Journal of Political Economy*, Vol. 81, No. 4: 950–961

Clark, James S., Steven R. Carpenter, Mary Barber, Scott Collins, Andy Dobson, Jonathan A. Foley, David M. Lodge, Mercedes Pascual, Roger Pielke Jr., William Pizer, Cathy Pringle, Walter V. Reid, Kenneth A. Rose, Osvaldo Sala, William H. Schlesinger, Diana H. Wall, David Wear. 2001. "Ecological Forecasts: An Emerging Imperative." *Science*, Vol. 293 (27 July 2001): 657–660

Clark, Peter U., Nicklas G. Pisias, Thomas F. Stocker and Andrew J. Weaver. 2002. "The role of the thermohaline circulation in abrupt climate change." *Nature*, Vol. 415 (21 February 2002): 863–869

Clarke, Tom. 2002a. "Wanted: scientists for sustainability." *Nature* (News feature), Vol. 418 (22 August 2002): 812–814

——. 2002b. "Exploitation costs the Earth." *Nature* (Science Update, August 9, 2002). [online] URL: www.nature.com/nsu/020805/020805-11.html

Cleeton, James. 2004. "Organic Foods in Relation to Nutrition and Health." In: *Coronary and Diabetic Care in the UK 2004*. Association of Primary Care Groups and Trust. Pp. 58–63

Clif Bar Inc. Our Story – Environment – Why Organic? [online] URL: www.clifbar.com/ourstory/document.cfm?location=environment&websubsection =organic

Cohen, Joel E., and David Tilman. 1996. "Biosphere 2 and Biodiversity – The Lessons So Far." *Science*, Vol. 274 (15 November 1996), No. 5290: 1050–1051

Colwell, Rita R., Anwar Huq, M. Sirajul Islam, K.M.A. Aziz, M. Yunus, N. Huda Khan, A. Mahmud, R. Bradley Sack, G.B. Nair, J. Chakraborty, David A. Sack, and E. Russek-Cohen. 2003. "Reduction of cholera in Bangladeshi villages by simple filtration." *PNAS*, Vol. 100, No. 3 (4 Feb. 2003): 1051–1055

Costa G. 1996. "The impact of shift and night work on health." *Ergonomics*, Vol. 27, No. 1 (1996): 6–16

Costanza, Robert; Ralph d'Arge, Rudolf de Groot, Stephen Farber, Monica Grasso, Bruce Hannon, Karin Limburg, Shahid Naeem, Robert V. O'Neill, Jose Paruelo, Robert G. Raskin, Paul Sutton & Marjan van den Belt. 1997. "The value of the world's ecosystem services and natural capital." *Nature*, Vol. 387 (15 May 1997): 253–260

Cox, Paul Alan. 1994. "Wild Plants as Food and Medicine in Polynesia." In: Etkin, Nina L. (ed.) *Eating on the Wild Side. The Pharmacologic, Ecologic, and Social Implications of Using Noncultigens.* Tucson & London: The University of Arizona Press. Pp. 102–113

Crumley, Carole (ed.), with A. Elizabeth van Deventer and Joseph J. Fletcher. 2001. *New Directions in Anthropology and Environment. Intersections.* Walnut Creek, CA: Altamira Press

Daily, Gretchen C. (ed.) 1997a. *Nature's Services. Societal Dependence on Natural Ecosystems.* Washington D.C. and Covelo, CA: Island Press

Daily, Gretchen C. 1997b. "Introduction: What are Ecosystem Services?" In: Daily, G.C. (ed.) *Nature's Services. Societal Dependence on Natural Ecosystems.* Washington D.C. and Covelo, CA.: Island Press. Pp. 1–10

Daily, Gretchen C. 2000. "Biodiversity and Happiness." In: Botkin et al. *Forces of Change. A New View of Nature.* Washington, D.C.: National Geographic Society. Pp. 227–237

Daily, Gretchen C., Pamela A. Matson, and Peter M. Vitousek. 1997. "Ecosystem Services Supplied by Soil." In: Daily, G.C. (ed.) *Nature's Services. Societal Dependence on Natural Ecosystems.* Washington D.C. and Covelo, CA.: Island Press. Pp. 113–132

Daly, Herman E. 1997. *Beyond Growth. The Economics of Sustainable Development.* Beacon Press

Daly, Herman E. and Joshua Farley. 2003. *Ecological Economics. Principles and Applications.* Island Press

Dasgupta, Partha. 2001. *Human Well-Being and the Natural Environment.* Oxford: Oxford University Press

Dasmann, Raymond F. 1976a. "Future primitive: ecosystem people versus biosphere people." *CoEvolution Quarterly* 11: 26–31

Dasmann, Raymond F. 1976b. "The Threatened World of Nature." XVI. The Horace M. Albright Conservation Lectureship (Berkeley, California; April 29, 1976). [online] URL:
www.cnr.berkeley.edu/forestry/lectures/albright/1976dasmann.html

Daily, Gretchen C., and Brian H. Walker. 2000. "Seeking the great transition." *Nature*, Vol. 403 (20 January 2000): 243–245

Dawkins, Richard. 1976. *The Selfish Gene.* Oxford: Oxford University Press

——. 1998. *Unweaving the Rainbow. Science, Delusion and the Appetite for Wonder.* London: Allen Lane (The Penguin Press)

de Bruyn, S.M. 1999. *Economic growth and the environment. An empirical analysis.* Amsterdam: Thela Thesis.

de Bruyn, S.M., and J.B. Opschoor. 1997. "Developments in the throughput-income relationship: Theoretical and empirical observations." *Ecological Economics*, Vol. 20: 255–268

Deffeyes, Kenneth S. 2003. *Hubbert's Peak. The Impending World Oil Shortage.* Princeton University Press

Dempster, W.F. 1991. "Biosphere II: engineering of manned, closed ecological systems." *Journal of Aerospace Engineering*, Vol. 4 (1): 23–30

De Young, Raymond. 2000. "Expanding and Evaluating Motives for Environmentally Responsible Behavior." *Journal of Social Issues*, Vol. 56, No. 3 (2000): 509–526

Diamond, Jared. 1992. *The Third Chimpanzee. The Evolution and Future of the Human Animal.* Harper Collins (Perennial)

——. 2005. *Collapse: How Societies Choose to Fail or Succeed.* Viking/Allen Lane

Dinges D.F. 1995. "An overview of sleepiness and accidents." *Journal of Sleep Research*, Vol. 4, Suppl. 2 (1995): 4–14

Douthwaite, R. 1992. *The Growth Illusion: How Economic Growth has Enriched the Few, Impoverished the Many, and Endangered the Planet.* Bideford, UK: Green Books

Dufour, Darna L. and Warren M. Wilson. 1994. "Characteristics of 'Wild' Plant Foods Used by Indigenous Populations in Amazonia." In: Etkin, Nina L. (ed.) *Eating on the Wild Side. The Pharmacologic, Ecologic, and Social Implications of Using Noncultigens.* Tucson & London: The University of Arizona Press. Pp. 114–142

Duke, James A. 2003. "Artemisia." *Herbal Gram, The Journal of the American Botanical Council* 57 (2003): 65

Durham, William H. 1991. *Coevolution. Genes, Culture, and Human Diversity.* Stanford, CA: Stanford University Press

Dusheck, Jennie. 2002. "The Interpretation of Genes." *Natural History* 10/02: 53–59

Eckersley, Richard. 2000. "The Mixed Blessings of Material Progress: Diminishing Returns in the Pursuit of Happiness." *Journal of Happiness Studies*, 1: 267–292

Egana, N. 2003. "Vitamin A Deficiency and Golden Rice – A Literature Review." *Journal of Nutritional & Environmental Medicine*, Vol. 13, No. 3 (September 2003): 169–184

Ekins, Paul. 2000. *Economic Growth and Environmental Sustainability. The Prospects for Green Growth.* London and New York: Routledge

Elgin, D. 1993. *Voluntary Simplicity. Toward a way of life that is outwardly simple, inwardly rich (revised edition).* New York: Quill

Ellen, Roy and Katsuyoshi Fukui (eds). 1996. *Redefining Nature. Ecology, Culture and Domestication.* Oxford, Washington, DC: Berg

EPI (Earth Policy Institute). 2004. Eco-Economy Update 36. [online] URL: www.earth-policy.org/Updates/Update36.htm

Esrey, Steven A., Jean Gough, Dave Rapaport, Ron Sawyer, Mayling Simpson-Hébert, Jorge Vargas, and Uno Winblad (ed.) 1998. *Ecological Sanitation.* Stockholm: SIDA

Esrey, Steven A. and Ingvar Andersson. 2001. *Ecological Sanitation. Closing the Loop.* UA Magazine (March 2001). [online] URL: http://www.gtz.de/ecosan/download/esrey2001.pdf

Etkin, Nina L. (ed.) 1994. *Eating on the Wild Side. The Pharmacologic, Ecologic, and Social Implications of Using Noncultigens.* Tucson & London: The University of Arizona Press

Etkin, Nina L. and Paul J. Ross. 1994. "Pharmacologic Implications of 'Wild' Plants in Hausa Diet." In: Etkin, Nina L. (ed.) *Eating on the Wild Side.* Tucson & London: The University of Arizona Press. Pp. 85–101

Evans, Peter (ed.) 2002. *Livable Cities? Urban Struggles for Livelihood and Sustainability.* Berkeley, Los Angeles, London: University of California Press

Evans, Peter. 2002a. "Introduction: Looking for Agents of Urban Livability in a Globalized Political Economy." In: Evans, Peter (ed.) *Livable Cities?* Berkeley, Los Angeles, London: University of California Press. Pp. 1–30

Ewald, Paul. W. 1994. *Evolution of Infectious Disease.* Oxford University Press

——. 1999a. "Using Evolution as a Tool for Controlling Infectious Diseases." In: Trevathan et al (eds). *Evolutionary Medicine.* Oxford University Press. Pp. 245–270

—— . 1999b. "Evolutionary Control of HIV and Other Sexually Transmitted Viruses." In: Trevathan et al (eds). *Evolutionary Medicine.* Oxford University Press. Pp. 271–312

Ewel, Katherine C. 1997. "Water Quality Improvement by Wetlands." In: Daily, G.C. (ed.) *Nature's Services. Societal Dependence on Natural Ecosystems.* Washington D.C. and Covelo, CA.: Island Press. Pp. 329–344

Faber Taylor, Andrea, F. Kuo and W.C. Sullivan. 2001. "Coping with ADD: The Surprising Connection to Green Play Settings." *Environment & Behavior,* Vol. 33, No. 1: 54–77

Faber Taylor, Andrea, Frances E. Kuo and William C. Sullivan. 2002. "Views of Nature and Self-Discipline. Evidence from Inner City Children." *Journal of Environmental Psychology,* 22: 49–63

Farnsworth, Norman R. 1988. "Screening Plants for New Medicines." In: Wilson, E.O. (ed.) *Biodiversity.* Washington, D.C.: National Academy of Sciences Press. Pp. 83–97

Fitzpatrick, Michael. 2000. *The Tyranny of Health. Doctors and the Regulation of Lifestyle.* London: Routledge

Foltz, Richard C., Frederick M. Denny, and Azizan Baharuddin (eds). 2003. *Islam and Ecology. A Bestowed Trust.* Cambridge, MA: Harvard University Press

Franklin, Jerry F. 1988. "Structural and Functional Diversity in Temperate Forests." In: Wilson, E.O. (ed.) *Biodiversity*. Washington, D.C.: National Academy of Sciences Press. Pp.166–175

Futuyma, Douglas, J. 1998. *Evolutionary Biology (Third Edition)*. Sunderland, MA: Sinauer Associates, Inc.

Gadgil, M. and R. Guha. 1992. *This Fissured Land: An Ecological History of India*. Oxford, New York: Oxford University Press

Gadgil, Madhav. 1993. "Of Life and Artifacts." In: Kellert, Stephen R. and E. O. Wilson (eds). *The Biophilia Hypothesis*. Washington, DC: Island Press. Pp. 365–377

Gandy, Matthew. 2003. *Concrete and Clay. Reworking Nature in New York City*. Cambridge, MA and London: The MIT Press

Gardner, Gary and Brian Halweil. 2000. "Overfed and Underfed. The Global Epidemic of Malnutrition" (Worldwatch Paper 150). Worldwatch Institute

Garwin, Laura and Ehsan Masood. 1998. "The worth of the Earth." *Nature* (Science Update, October 8, 1998). [online] URL: www.nature.com/nsu/981008/981008-3.html

George, S. 1993. "One-third in, two-thirds out." *New Perspectives Quarterly*, 10: 53–55

Girardot, N.J., James Miller and Liu Xiaogan (eds). 2001. *Daoism and Ecology. Ways Within a Cosmic Landscape*. Cambridge, MA: Harvard University Press

Glassner, Barry. 2000. *The Culture of Fear. Why Americans Are Afraid of the Wrong Things*. Basic Books

Gomez-Pompa, A. and A. Kaus. 1992. "Taming the wilderness myth." *BioScience*, Vol. 42: 271–279

Goodland, Robert. 2003. "Sustainability: Human, Social, Economic and Environmental." Article in: Ted Munn (ed.) *Encyclopedia of Global Environmental Change*. John Wiley & Sons, Ltd.

Gould, Stephen Jay. 1989. *Wonderful Life. The Burgess Shale and the Nature of History*. London: Penguin Books

Gould, Stephen J. 2001 [1999]. *Rocks of Ages. Science and Religion in the Fullness of Life*. London: Jonathan Cape

Grahn, P. and U.A. Stigsdotter. 2003. "Landscape planning and stress." *Urban Forestry & Urban Greening*, Vol. 2, No. 1 (June 2003): 1–18

Gray, John. 1998. *False Dawn. The Delusions of Global Capitalism*. London: Granta

Grim, John (ed.) 2001. *Indigenous Traditions and Ecology. The Interbeing of Cosmology and Community*. Cambridge, MA: Harvard University Press

Grubb, Michael. 2001. "Relying on Manna from Heaven?" (Science's Compass – Books) *Science*, Vol. 294 (9 November 2001): 1285–1287

Guha, R. and Martinez-Alier, J. 1997. *Varieties of Environmentalism. Essays North and South*. London: Earthscan

Gullone, Eleonora. 2000. "The Biophilia Hypothesis and Life in the 21st Century. Increasing mental health or increasing pathology?" *Journal of Happiness Studies*, Vol. 1 (2000): 293–321

Hall, C. A. S., C. J. Cleveland and R. Kaufmann. 1992. *Energy and Resource Quality: The Ecology of the Economic Process*. New York: Wiley Interscience

Hall, Peter. 1998. *Cities in Civilization. Culture, Innovation, and Urban Order.* London: Phoenix

Halweil, Brian. 2002. *Home Grown. The Case for Local Food in a Global Market.* Worldwatch Institute

Harmon, David. 2001. "On the Meaning and Moral Imperative of Diversity." In: Maffi, Luisa (ed.) *On Biocultural Diversity.* Washington and London: Smithsonian Institution Press. Pp. 53–70

Harr, R.D. 1982. "Fog drip in the Bull Run municipal watershed." *Oregon Water Res. Bull.* 18(5): 785–789

Harries-Jones, Peter. 1992. "Sustainable anthropology: ecology and anthropology in the future." In: Wallman, Sandra (ed.) *Contemporary Futures. Perspectives from Social Anthropology.* Pp. 157–171

Harris, Marvin. 1999. *Theories of Culture in Postmodern Times.* Walnut Creek, London, New Delhi: Altamira Press

Hartig, T., M. Mang and G.W. Evans. 1991. "Restorative effects of natural environment experiences." *Environment and Behavior* 23: 3–26

Hartig, T., G.W. Evans, L.D. Jamner, D.S. Davis and T. Gaerling. "Tracking restoration in natural and urban field settings." *Journal of Environmental Psychology,* Vol. 23, Issue 2 (June 2003): 109–123

Hatano, G. and K. Inagaki. 1987. "Everyday biology and school biology. How do they interact?" *The Quarterly Newsletter of the Laboratory of Human Cognition* 9: 120–128

Hawken, Paul, and Amory & Hunter Lovins. 1999. *Natural Capitalism.* Boston, MA: Little, Brown and Co.

Heckenberger, Michael J., Afukaka Kuikuro, Urissapá Tabata Kuikuro, J. Christian Russell, Morgan Schmidt, Carlos Fausto, and Bruna Franchetto. 2003. "Amazonia 1492: Pristine forest or cultural parkland?" *Science,* Vol. 301 (19 September 2003): 1710–1714

Heerwagen, Judith H. and Gordon H. Orians. 1993. "Humans, Habitats, and Aesthetics." In: Kellert, Stephen R. and E. O. Wilson (eds). *The Biophilia Hypothesis.* Washington, DC: Island Press. Pp. 138–172

Heerwagen, Judith H. and Gordon H. Orians. 2002. "The Ecological World of Children." In: Kahn, Peter H. Jr. and Stephen R. Kellert (eds). *Children and Nature.* Cambridge, MA and London: The MIT Press. Pp. 29–64

Heinberg, Richard. 2003. *The Party's Over. Oil, War, and the Fate of Industrial Societies.* Gabriola Island, B.C.: New Society Publishers

Herzog, Thomas R., Hong C. Chen and Jessica S. Primeau. 2002. "Perception of the restorative potential of natural and other settings." *Journal of Environmental Psychology,* 22: 295–306

Hessel, Dieter T. and Rosemary Radford Ruether (eds). 2000. *Christianity and Ecology. Seeking the Well-Being of Earth and Humans.* Cambridge, MA: Harvard University Press

Hewitt, Ben. 2003. "Harvest of Champions. If mom had told you what fruits and veggies can do for your game, maybe you would've listened. But it's not too late."

(Bodywork: The Performance Diet). *Outside Magazine* (March 2003). [online] URL: http://outside.away.com/outside/bodywork/200303/200303_bodywork_1.html

Hobson, K. 2002. "Competing Discourses of Sustainable Consumption: Does the 'Rationalisation of Lifestyles' Make Sense?" *Environmental Politics*, Vol. 11, No. 2 (Summer 2002): 95–120

Holling, C.S. 2000. "Theories for Sustainable Futures." *Conservation Ecology* 4(2): 7 [online] URL: www.consecol.org/vol4/iss2/art7

Honari, Morteza and Thomas Boleyn (eds). 1999. *Health Ecology. Health, Culture and Human-Environment Interaction.* London and New York: Routledge

Hooper, David U. and Jeffrey S. Dukes. 2004. "Overyielding among plant functional groups in a long-term experiment." *Ecology Letters* (2004), 7: 95–105

Hooper, Lora V. and Jeffrey I. Gordon. 2001. "Commensal Host-Bacterial Relationships in the Gut." *Science*, Vol. 292 (11 May 2001): 1115–1118

Hopfenberg, Russell, and David Pimentel. 2001. "Human Population Numbers as a Function of Food Supply." *Environment, Development, and Sustainability* 3: 1–15

Huber, K and N. Fuchs. 2003. "Wie wirkt die Erzeugungsqualität von Lebensmitteln? Ergebnisse der Ernährungs-Qualitäts-Studie des Forschungsring (Klosterstudie)." *Lebendige Erde* 4/2003: 42–47

Huxley, Aldous. 1944. *Time Must Have a Stop.* London: Chatto and Windus. Reprint 1994, London: Flamingo

Ilbery, B., Q. Chiotti, and T. Rickard (eds). 1997. *Agricultural Restructuring and Sustainability. A Geographic Perspective.* Wallingford, UK: CAB International

Iltis, Hugh H. 1988. "Serendipity in the Exploration of Biodiversity: What Good Are Weedy Tomatoes?" In: Wilson, E.O. (ed.) *Biodiversity.* Washington, D.C.: National Academy of Sciences Press. Pp. 98–105

Inagaki, K. 1990. "The effects of raising animals on children's biological knowledge." *British Journal of Developmental Psychology* 8: 119–129

Ingold, Tim. 1993. "Globes and Spheres. The topology of environmentalism." In: Milton, Kay (ed.) *Environmentalism: the view from anthropology.* London and New York: Routledge. Pp. 31–42

——. 2000. *The Perception of the Environment. Essays on Livelihood, Dwelling and Skill.* London and New York: Routledge

Jackson, L.E. (ed.) 1997. *Ecology in Agriculture.* San Diego: Academic

Jackson, Dana L. and Laura L. Jackson. 2002. *The Farm as Natural Habitat. Reconnecting Food Systems with Ecosystems.* Washington, Covelo, London: Island Press

Jenkins, Martin. 2003. "Prospects for Biodiversity." *Science* (State of the Planet Special Series), Vol. 302 (14 November 2003): 1175–1177

Johns, Timothy. 1990. *With Bitter Herbs They Shall Eat It. Chemical Ecology and the Origins of Human Diet and Medicine.* Tucson: University of Arizona Press

Johns, Timothy. 1999. "Plant Constituents and the Nutrition and Health of Indigenous Peoples." In: Nazarea, Virginia D. (ed.) *Ethnoecology. Situated Knowledge/Located Lives.* Tucson: The University of Arizona Press. Pp. 157–174

Jordan, Nicholas R. 2002. "Sustaining Production with Biodiversity." In: Jackson, Dana L. and Laura L. Jackson. *The Farm as Natural Habitat. Reconnecting Food Systems with Ecosystems.* Washington, Covelo, London: Island Press. Pp. 155–168

Jordan, William R., III. 1988. "Ecological Restoration: Reflections on a Half-Century of Experience at the University of Wisconsin-Madison Arboretum." In: Wilson, E.O. (ed.) 1988. *Biodiversity.* Washington, D.C.: National Academy Press. Pp. 311–316

Josephson, Paul R. 2002. *Industrialized Nature. Brute Force Technology and the Transformation of the Natural World.* Washington, D.C.: Island Press

Kabesch, Michael and Roger P. Lauener. 2004. "Why Old McDonald had a farm but no allergies: genes, environments, and the hygiene hypothesis." *Journal of Leukocyte Biology* 75: 383–387

Kahn, Peter H., Jr. 1999. *The Human Relationship with Nature. Development and Culture.* Cambridge, MA and London: The MIT Press

Kahn, Peter H., Jr. and Stephen R. Kellert (eds). 2002. *Children and Nature. Psychological, Sociocultural, and Evolutionary Investigations.* Cambridge, MA, and London: The MIT Press

Kaiser, Jocelyn. 2001. "The Other Global Pollutant. Nitrogen proves tough to curb." *Science*, Vol. 294 (9 Nov. 2001): 1268–1269

Kates, Robert W.; Clark, William C.; et al. 2001. "Sustainability Science." *Science* (Science's Compass – Policy Forum: Environment and Development), Vol. 292, No. 5517 (27 April 2001): 641–642

Kaplan, R. 1973. "Some psychological benefits of gardening." *Environment and Behavior* 5: 145–152

——. 1974. "Some psychological benefits of an outdoor challenge program." *Environment and Behavior* 6: 101–116

——. 1977. "Patterns of environmental preference." *Environment and Behavior* 9: 195–216

——. 1993. "The role of nature in the context of the workplace." *Landscape and Urban Planning* 26: 193–201

——. 2001. "The nature of the view from home: psychological benefits." *Environment & Behavior*, Vol. 33, No. 4 (July 2001): 507–542

Kaplan, R. and S. Kaplan. 1989. *The Experience of Nature. A Psychological Perspective.* New York: Cambridge University Press

Kaplan, Stephen. 2000. "Human Nature and Environmentally Responsible Behavior." *Journal of Social Issues*, Vol. 56, No. 3 (2000): 491–508

Kasser, Tim and Allen D. Kanner (eds). 2004. *Psychology and Consumer Culture. The Struggle for a Good Life in a Materialistic World.* Washington, DC: American Psychological Association

Kay, James J., Henry A. Regier, Michelle Boyle, and George Francis. 1999. "An ecosystem approach for sustainability: addressing the challenge of complexity." *Futures* 31 (1999): 721–742

Kellert, Stephen R. 1993. "The Biological Basis for Human Values of Nature." In: Kellert, Stephen R. and E. O. Wilson (eds). *The Biophilia Hypothesis*. Washington, DC: Island Press. Pp. 42–69

Kellert, Stephen R. and E. O. Wilson (eds). 1993. *The Biophilia Hypothesis*. Washington, DC: Island Press

Kellert, Stephen R. 1996. *The Value of Life. Biological Diversity and Human Society*. Washington, DC: Island Press

——. 1997. *Kinship to Mastery. Biophilia in Human Evolution and Development*. Washington, DC: Island Press

Kimbrell, Andrew. 2002. *Fatal Harvest. The Tragedy of Industrial Agriculture*. Island Press

Kingsnorth, Paul. 2003. *One No, Many Yeses: A Journey to the Heart of the Global Resistance Movement*. Free Press/Simon & Schuster

Kinsley, David. 1995. *Ecology and Religion. Ecological Spirituality in Cross-Cultural Perspective*. Englewood Cliffs, NJ: Prentice Hall

Kirchner, James W. 2002. "Evolutionary speed limits inferred from the fossil record." *Nature*, Vol. 415 (3 January 2002): 65–67

Kirchner, James W. and Anne Weil. 2000a. "Correlations in fossil extinction and origination rates through geological time." *Proceedings of the Royal Society London B* (2000) 267, 1301–1309

Kirchner, James W. and Anne Weil. 2000b. "Delayed biological recovery from extinctions throughout the fossil record." *Nature*, Vol. 404 (9 March 2000): 177–180

Klare, Michael T. 2002. *Resource Wars: The New Landscape of Global Conflict (Reprint Edition with new introduction)*. Owl Books

Kluger, Jeffrey and Andrea Dorfman, 2002. "The challenges we face." *TIME Magazine* (Special Report), August 2002

Korzybski, Alfred. 1958 [1933]. *Science and Sanity*. 4th ed. The International Non-Aristotelian Library

Kuckartz, Udo, 2000. *Umweltbewusstsein in Deutschland 2000. Ergebnisse einer repräsentativen Bevölkerungsumfrage (Stand: Juni 2000)*. Berlin: Bundesministerium für Umwelt, Naturschutz und Reaktorsicherheit – Referat Gesellschaftspolitische Grundsatzfragen

Kutas, M. and K.D. Federmeier. 1998. "Minding the body." *Psychophysiology*, Vol. 35, No. 2 (March 1998): 135–150

Lagrotteria, Marta and James M. Affolter. 1999. "Sustainable Production and Harvest of Medicinal and Aromatic Herbs in the Sierras de Córdoba Region, Argentina." In: Nazarea, Virginia D. (ed.) *Ethnoecology. Situated Knowledge/ Located Lives*. Tucson: University of Arizona Press. Pp. 175–189

Langer, Ellen. 1983. *The Psychology of Control*. Beverly Hills: Sage

Lappé, Frances Moore. 1991 [1971]. *Diet for a Small Planet (20th Anniversary Edition)*. New York: Ballantine Books

Lappé, Frances Moore, Joseph Collins, and Peter Rosset, with Luis Esparza. 1998. *World Hunger. Twelve Myths. 2nd ed.* (The Institute for Food and Development Policy). New York: Grove Press

Lappé, Frances Moore and Anna Lappé. 2002. *Hope's Edge.* Tarcher/Putnam

Lawrence, Elizabeth Atwood. 1993. "The Sacred Bee, the Filthy Pig, and the Bat Out of Hell: Animal Symbolism as Cognitive Biophilia." In: Kellert, Stephen R. and E. O. Wilson (eds). *The Biophilia Hypothesis.* Washington, DC: Island Press. Pp. 301–341

Leger, D. 1994. "The cost of sleep-related accidents: A report for national commission on sleep disorders." *Sleep*, Vol.17, No. 1 (1994): 84–93

Leopold, Aldo. 1986 [1949]. *A Sand County Almanac. Reissue* (copyright 1966 Oxford University Press). Ballantine Books

Lessig, Lawrence. 2004. "Insanely Destructive Devices. Trying to defend against self-replicating weapons of mass destruction." *Wired*, Issue 12.04 (April 2004) [online] URL: www.wired.com/wired/archive/12.04/

Li, Wenhua (ed.) 2001. *Agro-Ecological Farming Systems in China* (Man and the Biosphere Series, Volume 26). Paris: UNESCO, and New York, Lancs: The Parthenon Publishing Group

Lomborg, Bjorn. 2001. *The Skeptical Environmentalist. Measuring the Real State of the World.* Cambridge: Cambridge University Press

Lovins, Amory B.; L. Hunter Lovins, and Paul Hawken. 1999. "A Road Map For Natural Capitalism." *Harvard Business Review*, May-June 1999: 145–158

Lovins, Amory B., and L. Hunter Lovins. 2001. "Natural Capitalism: Path to Sustainability?" *Corporate Environmental Strategy*, Vol. 8, No. 2 (2001): 99–108

Lovins, Amory B, E. Kyle Datta, Thomas Feiler, Karl R. Rábago, Joel N. Swisher P.E., André Lehmann, and Ken Wicker. 2002. *Small is Profitable. The Hidden Economic Benefits of Making Electrical Resources the Right Size.* Rocky Mountain Institute

Mackenzie, Fred T. 2003. *Our Changing Planet. An Introduction to Earth System Science and Global Environmental Change (Third Edition).* Upper Saddle River, NJ: Pearson Education, Inc. (Prentice Hall)

MacMahon, James A. and Karen D. Holl. 2001. "Ecological Restoration. A Key to Conservation Biology's Future." In: Soulé & Orians. *Conservation Biology.* Washington, D.C.: Island Press

Maeder, Paul, Andreas Fliessbach, David Dubois, Lucie Gunst, Padruot Fried and Urs Niggli. 2002. "Soil Fertility and Biodiversity in Organic Farming." *Science*, Vol. 296 (31 May 2002): 1694–1697

Maffi, Luisa (ed.) 2001. *On Biocultural Diversity. Linking language, knowledge, and the environment.* Washington and London: Smithsonian Institution Press

Maquet, P. 2001. "The role of sleep in learning and memory." *Science*, Vol. 294: 1048–1052

Marten, Gerald G. 2001. *Human Ecology. Basic Concepts for Sustainable Development.* London: Earthscan

Mascie-Taylor, C.G. and Enamul Karim. 2003. "The Burden of Chronic Disease." *Science*, Vol. 302 (12 December 2003): 1921–1922

Maslow, Abraham. 1970. *Motivation and Personality, 2nd ed.* New York: Harper and Row

McDonough, William, and Michael Braungart. 2002. *Cradle to Cradle. Remaking the Way We Make Things.* New York: North Point Press

McMichael, Tony (Anthony J). 2001. Hu*man Frontiers, Environments and Disease. Past Patterns, Uncertain Futures.* Cambridge: Cambridge University Press

——. 2002a. "The Biosphere, Health, and 'Sustainability'." *Science*, Vol. 297 (16 August 2002), No. 5584: 1093

——. 2002b. "Fine Battlefield Reporting, But It's Time to Stop the War Metaphor." *Science*, Vol. 295 (22 February 2002): 1469

——. 2002c. "Population, environment, disease, and survival: past patterns, uncertain futures." *The Lancet*, Vol. 359 (30 March 2002): 1145–48

McNeely, J.A. 1989. "Protected Areas and Human Ecology: How National Parks Can Contribute to Sustaining Societies of the Twenty-First Century." In: D. Western and M. Pearl (eds). *Conservation for the Twenty-first Century.* Oxford, New York: Oxford University Press Pp. 150–165

McNeely, Jeffrey A. and Sandra Scherr. 2001. *Common Ground, Common Future. How Ecoagriculture Can Help Feed The World And Save Wild Biodiversity.* IUCN and Future Harvest

Meher-Homji, V.M. 1992. "Probable Impact of Deforestation on Hydrological Processes." In: Myers, N. (ed.) Tropical Forests and Climate. *Climatic Change* 19:1–2 (special issue): 163–174. Dordrecht, Netherlands: Kluwer Academic Publisher.

Moavenzadeh, Fred, Keisuke Hanaki, and Peter Baccini. 2002. Future Cities: Dynamics and Sustainability. Kluwer Academic Publishers

Moerman, Daniel E. 1994. "North American Food and Drug Plants." In: Etkin, Nina L. (ed.) *Eating on the Wild Side.* Tucson & London: The University of Arizona Press. Pp. 166–181

Moffat, Anne Simon. 1998. "Global Nitrogen Overload Problem Grows Critical." *Science*, Vol. 279, No. 5353 (13 Feb. 1998): 988–989

Moran, Emilio F. 1996. "Nurturing the Forest. Strategies Among Native Amazonians." In: Ellen, Roy and Katsuyoshi Fukui (eds). *Redefining Nature.* Oxford, Washington, DC: Berg. Pp. 531–555

——. 2000. *Human Adaptability. An Introduction to Ecological Anthropology (Second Edition).* Boulder, CO, Oxford: Westview Press

Moran, Katy. 1999. "Toward Compensation. Returning Benefits from Ethnobotanical Drug Discovery to Native Peoples." In: Nazarea, Virginia D. (ed.) *Ethnoecology. Situated Knowledge/ Located Lives.* Tucson: University of Arizona Press. Pp. 249–262

Mueller, M.S., I.B. Karhagomba, H.M. Hirt and E. Wemakor. 2000. "The potential of Artemisia annua L. as a locally produced remedy for malaria in the tropics: agricultural, chemical and clinical aspects." *Journal of Ethnopharmacology*, Vol. 73, Issue 3 (December 2000): 487–493

Mulder, Peter, and Jeroen C.J.M. Van Den Bergh. 2001. "Evolutionary Economic Theories of Sustainability." *Growth and Change*, Vol. 32 (Winter 2001): 110–134

Munich Re (Muenchener Rueckversicherungsgesellschaft, Corporate Underwriting/Global Clients, GeoRisks Research Dept). 2004. *TOPICSgeo. Annual Review. Natural Catastrophes 2003*

Murphy, Dennis D. 1986. "Challenges to Urban Biological Diversity in Urban Areas." In: Wilson, E.O. (ed.) *Biodiversity*. Washington, D.C.: National Academy of Sciences Press. Pp. 71–76

Murray, Geoffrey and Ian G. Cook. 2002. *Green China. Seeking Ecological Alternatives*. London and New York: RoutledgeCurzon

Myers, J.P., L.J. Guillette, Jr., P. Palanza, S. Parmigiani, S.H. Swan, and F.S. vom Saal. 2004. "The emerging science of endocrine disruption." In: R. Ragaini (ed.) *International Seminar on Nuclear War and Planetary Emergencies – 30th Session* (The Science and Culture Series). World Scientific Publishing Co.

Myers, Norman (ed.) 1992. "Tropical Forests and Climate." *Climatic Change* 19:1–2 (special issue). Dordrecht, Netherlands: Kluwer Academic Publisher

Myers, Norman and J. Nancy. 1993. *Ultimate Security. The Environmental Basis of Political Stability*. W.W. Norton & Company

Myers, Norman. 1997. "The World's Forests and Their Ecosystem Services." In: Daily, G.C. (ed.) *Nature's Services. Societal Dependence on Natural* Ecosystems. Washington D.C. and Covelo, CA.: Island Press. Pp. 215–236

Myers, Norman and Jennifer Kent. 2001. *Perverse Subsidies*. Washington, D.C.: Island Press

Nabhan, Gary Paul. 2004. *Why Some Like It Hot. Food, Genes, and Cultural Diversity*. Washington, Covelo, London: Island Press / Shearwater Books

Nabhan, Gary Paul and Stephen L. Buchmann. 1997. "Services Provided by Pollinators." In: Daily, G.C. (ed.) *Nature's Services. Societal Dependence on Natural Ecosystems*. Washington D.C. and Covelo, CA.: Island Press. Pp. 133–150

Nations, James D., 2001. "Indigenous Peoples and Conservation. Misguided Myths in the Maya Tropical Forest." In: Maffi, Luisa. *On Biocultural Diversity. Linking language, knowledge, and the environment*. Washington and London: Smithsonian Institution Press. Pp. 462–471

Nature (no author given), 20 March 2003. "How to slake a planet's thirst" (Editorial). Vol. 422, Issue No. 6929: 243

Naylor, Rosamond L. and Paul R. Ehrlich. 1997. "Natural Pest Control Services and Agriculture." In: Daily, G.C. (ed.) *Nature's Services. Societal Dependence on Natural Ecosystems*. Washington D.C. and Covelo, CA.: Island Press. Pp. 151–175

Nazarea, Virginia D. 1998. *Cultural Memory and Biodiversity*. Tucson: University of Arizona Press

Nazarea, Virginia D (ed.) 1999. *Ethnoecology. Situated Knowledge/ Located Lives*. Tucson: University of Arizona Press

NEETF (The National Environmental Education & Training Foundation). 1999. *Environmental Readiness for the 21st Century. The Eighth Annual National Report Card on Environmental Attitudes, Knowledge, and Behavior (December 1999)*. Washington, D.C.: NEETF / Roper Starch Worldwide

Nestle, Marion. 2003. The Ironic Politics of Obesity (Editorial). Science, Vol. 299 (7 February 2003): 781

Nietschmann, B.Q. 1992. "The Interdependence of Biological and Cultural Diversity." *Occasional Paper* no. 21 (December 1992), Center for World Indigenous Studies

Nisbett, Richard E. 2003. *The Geography of Thought. How Asians and Westerners Think Differently … And Why*. New York: The Free Press (Simon & Schuster Inc.)

Nosengo, Nicola. 2003. "Fertilized to death. Vast quantities of nitrogen being poured onto farmer's fields are wreaking havoc with our forests" (News feature). *Nature*, Vol. 425 (30 Oct. 2003): 894–895

O'dea, K. 1984. "Marked Improvement in Carbohydrate and Lipid Metabolism in Diabetic Australian Aborigines after Temporary Reversion to Traditional Lifestyle." *Diabetes* 33: 596–603

Odling-Smee, John, Kevin N. Laland, and Marcus W. Feldman. 2003. *Niche Construction. The Neglected Process in Evolution*. Princeton University Press

Odum, H.T. 1971. *Environment, Power and Society*. New York: Wiley-Interscience

O'Meara, Molly. 1999. *Reinventing Cities for People and the Planet* (Worldwatch Paper 147). Worldwatch Institute

Opotow, Susan and Leah Weiss. 2000. "Denial and the Process of Moral Exclusion in Environmental Conflict." *Journal of Social Issues*, Vol. 56, No. 3 (2000): 475–490

Orr, David W. 2002. The Nature of Design. Ecology, Culture, and Human Intention. Oxford, New York: Oxford University Press

Oskamp, Stuart. 2000. "Psychological Contributions to Achieving an Ecologically Sustainable Future for Humanity." *Journal of Social Issues*, Vol. 56, No. 3 (2000): 373–390

Pacini, Cesare, Ada Wossink, Gerard Giesen, Concetta Vazzana and Ruud Huirne. 2003. "Evaluation of sustainability of organic, integrated and conventional farming systems: a farm and field-scale analysis." *Agriculture, Ecosystems and Environment*, Vol. 95 (2003): 273–288

Pandey, Deep Narayan, Anil K. Gupta, and David M. Anderson, 2003. "Rainwater harvesting as an adaptation to climate change" (Review Article). *Current Science*, Vol. 85, No. 1 (10 July 2003): 46–59

Parrot, Nicholas and Terry Marsden. 2002. *The Real Green Revolution. Organic and Agroecological Farming in the South*. London: Greenpeace Environmental Trust

Parsons, R., L.G. Tassinary, R.S. Ulrich, M.R. Hebl, and M. Grossman-Alexander. 1998. "The view from the road: Implications for stress recovery and immunization." *Journal of Environmental Psychology* 18: 113–140

Peet, Richard and Michael Watts (eds). 1996. *Liberation Ecologies. Environment, Development, Social Movements*. London: Routledge

Peña, Devon G. 1999. "Cultural Landscapes and Biodiversity. The Ethnoecology of an Upper Rio Grande Watershed Commons." In: Nazarea, Virginia D. (ed.) *Ethnoecology. Situated Knowledge/Located Lives*. Tucson: The University of Arizona Press. Pp. 107–132

Peters, R.H. 1991. *A Critique for Ecology*. Cambridge: Cambridge University Press

Peterson, Christopher, Steven F. Maier, and Martin E.P. Seligman. 1995. *Learned Helplessness. A Theory for the Age of Personal Control*. Oxford, New York: Oxford University Press

Pierskalla, Chad D., Martha E. Lee, Taylor V. Stein, Dorothy H. Anderson, and Ron Nickerson. 2004. "Understanding relationships among recreation opportunities: a meta-analysis of nine studies." *Leisure Sciences*, Vol. 26, No. 2 (April-June 2004): 163–180

Pimentel, D., J. Allen, A. Beers, L. Guinand, A. Hawkins, R. Linder, P. McLaughlin, B. Meer, D. Musonda, D. Perdue, S. Poisson, R. Salazar, S. Siebert, and K. Stoner. 1993. "Soil Erosion and Agricultural Productivity." In: Pimentel, D. (ed.) *World Soil Erosion and Conservation*. Cambridge, England: Cambridge University Press Pp. 277–292

Pinker, Steven. 2002. *The Blank Slate. The Modern Denial of Human* Nature. Penguin Putnam

Pirages, Dennis Clark and Theresa Manley DeGeest. 2004. *Ecological Security. An Evolutionary Perspective on Globalization*. Lanham, Boulder, New York, Toronto, Oxford: Rowman & Littlefield Publishers, Inc.

Plotkin, Mark J. 1988. "The Outlook for New Agricultural and Industrial Products from the Tropics." In: Wilson, E.O. (ed.) *Biodiversity*. Washington, D.C.: National Academy of Sciences Press. Pp. 106–116

Ponting, Clive. 1991. *A Green History of the World. The Environment and the Collapse of Great Civilizations*. London (et al.): Penguin

Posey, Darrell. A. (ed.) 1999. *Cultural and Spiritual Values of Biodiversity*. London: Intermediate Technology Publications for the United Nations Environment Programme, Kenya

Posey, Darrell A. 2001. "Biological and Cultural Diversity: The Inextricable, Linked by Language and Politics." In: Maffi, Luisa. *On Biocultural Diversity*. Washington and London: Smithsonian Institution Press. Pp. 379–396

Posey, Darrell A. and Graham Dutfield (eds). 1996. *Indigenous Peoples and Sustainability: Cases and Actions*. Gland and Utrecht: IUCN and International Books

Postel, Sandra. 1997. *Pillar of Sand. Can the Irrigation Miracle Last?* New York: W.W. Norton and Company

Postel, Sandra and Stephen Carpenter. 1997. "Freshwater Ecosystem Services." In: Daily, G.C. (ed.) *Nature's Services*. Societal Dependence on Natural Ecosystems. Washington D.C. and Covelo, CA.: Island Press. Pp. 195–214

Preston, Samuel. 1994. *Population and the Environment*. Liége, Belgium: International Union for the Scientific Study of Population (Distinguished Lecture Series on Population and Development)

Primack, Richard B. 1998. *Essentials of Conservation Biology (2nd. ed.)*. Sunderland, MA: Sinauer Associates

Proschan, Frank. 2003. *Vietnam's Ethnic Mosaic*. In: Va Huy, Nguyen, and Lauren Kendall (eds). Vietnam: Journeys of Body, Mind, and Spirit (Companion Publication to AMNH exhibition). Berkeley, Los Angeles: University of California Press. Pp. 52–69

Rammel, Christian and Jeroen C.J.M. van den Bergh. 2003. "Evolutionary policies for sustainable development: adaptive flexibility and risk minimising." *Ecological Economics*, 47 (2003): 121–133

Rappaport, Roy. 1984. *Pigs for the Ancestors: Ritual in the Ecology of a New Guinea People. A New, Enlarged Edition*. New Haven: Yale University Press

Rawls, John F., Buck S. Samuel, Jeffrey I. Gordon. 2004. "Gnotobiotic zebrafish reveal evolutionarily conserved responses to the gut microbiota." *PNAS* 10 (March 19, 2004): 1073

Ray, Paul H. and Sherry Ruth Anderson. 2000. *The Cultural Creatives. How 50 Million People are Changing the World.* New York: Three Rivers Press

Rees, Martin. 2003. *Our Final Century. Will the Human Race Survive the Twenty-First Century?* Heinemann

Rees, William E. 2002. "Footprint: our impact on Earth is getting heavier." *Nature* (Correspondence), Vol. 420 (21 November 2002): 267f.

Renner, Michael. 1996. *Fighting for Survival. Environmental Decline, Social Conflict, and the New Age of Insecurity.* W.W. Norton

——. 2002. *The Anatomy of Resource Wars.* Worldwatch Institute

Rhoades, Robert E. (ed.) 2001. *Bridging Human and Ecological Landscapes: Participatory Research and Sustainable Development in an Andean Frontier.* Dubuque: Kendall/Hunt Publishing Company

Rhoades, Robert E. and Jack Harlan, 1999. "Epilogue: Quo Vadis? The Promise of Ethnoecology." In: Nazarea, Virginia D. (ed.) *Ethnoecology. Situated Knowledge/ Located Lives.* Tucson: The University of Arizona Press Pp. 271–280

Richards, Paul. 1993. "Cultivation: Knowledge or Performance?" In: Hobart, Mark (ed.) *An Anthropological Critique of Development.* London: Routledge

Riedler J, Braun-Fahrlander C, Eder W, Schreuer M, Waser M, Maisch S, Carr D, Schierl R, Nowak D, von Mutius E; ALEX Study Team. 2001. "Exposure to farming in early life and development of asthma and allergy: a cross-sectional survey." *Lancet*, Vol. 358, No. 9288 (6 October 2001): 1129–33

Rocky Mountain Institute; Alex Wilson, Jenifer L. Uncapher, Lisa McManigal, L. Hunter Lovins, Maureen Cureton, William D. Browning. 1998. *Green Development: Integrating Ecology and Real Estate.* John Wiley & Sons

Rojzman, Charles. 2002. See: www.arte-tv.com/dossier/archive.jsp?refresh=false &node=14271&lang=fr (or &lang=de)

Rosenzweig, Michael L. 2003. *Win-Win Ecology. How the Earth's Species Can Survive In The Midst of Human Enterprise.* Oxford, New York: Oxford University Press

Rozin, Elisabeth. 1992 [1983]. *Ethnic Cuisine.* New York, London: Penguin Books (originally published 1983, Lexington, Mass.: S. Greene Press)

Sabloff, Jeremy A. 1990. *The New Archaeology and the Ancient Maya.* New York: Scientific American Library

Sagan, Dorion and Lynn Margulis. 1993. "God, Gaia, and Biophilia." In: Kellert, Stephen R. and E. O. Wilson (eds). *The Biophilia Hypothesis.* Washington, DC: Island Press. Pp. 345–364

Salati, E. and C.A. Nobre. 1992. "Possible Climatic Impacts of Tropical Deforestation." In: Myers, N. (ed.) Tropical Forests and Climate. *Climatic Change* 19:1–2 (special issue). Dordrecht, Netherlands: Kluwer Academic Publisher. Pp. 177–196

Sandell, Klas. 1995. "Nature as the Virgin Forest. Farmers' Perspectives on Nature and Sustainability in Low-Resource Agriculture in the Dry Zone of Sri Lanka." In:

Bruun, Ole and Arne Kalland (eds). *Asian Perceptions of Nature*. Richmond: Curzon Press. Pp. 148–172

Sanderson, Eric W., Malanding Jaiteh, Marc A. Levy, Kent H. Redford, Antoinette V. Wannebo, and Gillian Woolmer. 2002. "The Human Footprint and the Last of the Wild." *BioScience*, Vol. 52 (October 2002), No. 10: 891–904

Scheffer, Marten, Steve Carpenter, Jonathan A. Foley, Carl Folke, and Brian Walker. "Catastrophic Shifts in Ecosystems (Review Article)." *Nature*, Vol. 413 (11 October 2001): 591–596

Schellnhuber, H.-J. and V. Wenzel (eds). 1998. *Earth System Analysis: Integrating Science for Sustainability*. Berlin: Springer

Schettler, Ted, Gina Solomon, Maria Valenti, Annette Huddle. 2000. *Generations at Risk. Reproductive Health and the Environment*. Cambridge, MA and London: The MIT Press

Schneider, Stephen H. and Randi Londer. 1984. *The Coevolution of Climate and Life*. Sierra Club Books

Schulz, Richard. 1976. "Some Life and Death Consequences of Perceived Control." In: John S. Carroll and John W. Payne (eds). *Cognition and Social Behavior*. New York: Academic Press. Pp. 135–153

Schwartz, Peter and Doug Randall. 2003. *An Abrupt Climate Change Scenario and Its Implications for United States National Security, October 2003*

Schwartz-Nobel, Loretta. 2002. *Growing Up Empty: The Hunger Epidemic in America*. HarperCollins

Seligman, Martin E. P. and Mihaly Csikszentmihaly, 2000. "Positive Psychology: An Introduction." [online] URL: www.psych.upenn.edu/seligman/intro.htm Published in: American Psychologist, Vol. 55 (January 2000; Special Issue on Positive Psychology), No. 1

Selman, Paul. 2000. "Landscape Sustainability at the National and Regional Scales." In: Benson, John F. & Maggie H. Roe (eds). *Landscape and Sustainability*. London and New York: Spon Press. Pp. 97–110

Seo, Danny. 2001. *Conscious Style Home. Eco-Friendly Living for the 21st Century*. St. Martin's Press

Shanahan, James and Katherine McComas, 1999. *Nature Stories. Depictions of the Environment and their Effects* (The Hampton Press Communication Series). Cresskill, NJ: Hampton Press, Inc.

Sheets, V.L. and C.D. Manzer. 1991. "Affect, cognition, and urban vegetation: Some effects of adding trees along city streets." *Environment and Behavior* 23: 285–304

Shellenberger, Michael and Ted Nordhaus. 2005. "The Death of Environmentalism. Global Warming Politics in a Post-Environmental World." [online] URL: www.grist.org/news/maindish/2005/01/13/doe-reprint/

Shepard, Paul. 1993. On Animal Friends. In: Kellert, Stephen R. and E. O. Wilson (eds). The Biophilia Hypothesis. Washington, DC: Island Press. Pp. 275–300

Sherman, Paul W. and Jennifer Billing. 1999. "Darwinian Gastronomy: Why we use spices. Spices taste good because they are good for us." *BioScience* Vol. 49, No. 6 (June 1999): 453–463

Sherman, Paul W. and Samuel M. Flaxman. 2001. "Protecting Ourselves from Food. Spices and morning sickness may shield us from toxins and microorganisms in the diet." *American Scientist*, Vol. 89, No. 2 (March-April 2001): 142ff.

Shiklomanov, I.A. 1993. "World Freshwater Resources." In: Gleick, P.H. (ed.) *Water in Crisis. A Guide to the World's Freshwater Resources*. Oxford, New York: Oxford University Press

Shintani, T.T., C.K. Hughes, S. Beckham, and H.K. O'Connor. 1991. "Obesity and Cardiovascular Risk Intervention through the Ad Libitum Feeding of Traditional Hawaiian Diet." *American Journal of Clinical Nutrition* 53: 1647–1651

Simmons, I.G. 1997. *Humanity and Environment. A Cultural Ecology*. Essex: Addison Wesley Longman

Singer, Peter. 2004. *One World. The Ethics of Globalization (2nd edition)*. Yale University Press

Smail, J. Kenneth. 2002. "Confronting A Surfeit of People: Reducing Global Human Numbers to Sustainable Levels. An essay on population two centuries after Malthus." *Environment, Development, and Sustainability*, Vol. 4: 21–50

Smil, Vaclav. 1994. *Global Ecology. Environmental Change and Social Flexibility*. London, New York: Routledge

——. 2003. *Energy at the Crossroads. Global Perspectives and Uncertainties*. Cambridge, MA and London: MIT Press

Snow, C.P. 1959. *The Two Cultures* (The Rede Lecture). Part I in reissue: 1998. The Two Cultures. Cambridge, New York: Cambridge University Press

Soil Association – Heaton Shane. 2001. *Organic Farming, Food Quality and Human Health. A review of the evidence*

Soleri, Daniela and Steven E. Smith. 1999. "Conserving Folk Crop Varieties. Different Agricultures, Different Goals." In: Nazarea, Virginia D. (ed.) *Ethnoecology. Situated Knowledge/Located Lives*. Tucson: The University of Arizona Press. Pp. 133–154

Soulé, Michael E. and Gordon H. Orians (eds). 2001. *Conservation Biology. Research Priorities for the Next Decade*. Washington, D.C.: Island Press

Soulé, Michael E. and Gordon H. Orians. 2001. "Conservation Biology Research. Its Challenges and Contexts." In: Soulé, Michael E. and Gordon H. Orians (eds). *Conservation Biology*. Washington, D.C.: Island Press. Pp. 271–285

Sperber, Birgitte Glavind. 1995. "Nature in the Kalasha Perception of Life." In: Bruun, Ole and Arne Kalland (eds). *Asian Perceptions of Nature*. Richmond: Curzon Press. Pp. 126–147

Speth, James Gustave. 2004. *Red Sky at Morning. America and the Crisis of the Global Environment*. New Haven and London: Yale University Press

Sponsel, Leslie E. 2001. "Do Anthropologists Need Religion, and Vice Versa? Adventures and Dangers in Spiritual Ecology." In: Crumley, Carole (ed.) *New Directions in Anthropology and Environment. Intersections*. Walnut Creek, CA: Altamira Press. Pp. 177–200

Stearns, Stephen C. (ed.) 1999. *Evolution in Health and Disease*. Oxford: Oxford University Press

Strange, Mark. 1988 [1999]. "The Faustian Bargain: Technology and the Price Issue." In: Boucher, Douglas H. (ed.) *The Paradox of Plenty*. Pp. 163–171. [Originally published in: Strange, Mark. 1988. Family Farming. A New Economic Vision. University of Nebraska Press]

Sullivan, Lawrence E. 1999. "Preface." In: Tucker, Mary Evelyn and Duncan Ryuken Williams (eds). *Buddhism and Ecology. The Interconnection of Dharma and Deeds*. Cambridge, MA: Harvard University Press. Pp. xi–xiv

Teisl, Mario F. and Kelly O'Brien. 2003. "Who cares and who acts? Outdoor recreationists exhibit different levels of environmental concerns and behavior." *Environment & Behavior*, Vol. 35, No. 4 (July 2003): 506–522

Tennessen, C. M. and B. Cimprich. 1995. "View to nature: effects on attention." *Journal of Environmental Psychology*, Vol. 15: 77–85

Thaman, Konai H. 2002. "Shifting sights. The cultural challenge of sustainability." *Journal of Sustainability in Higher Education*, Vol. 3, No. 3 (2002): 233–242

Thiele, Leslie Paul. 2001. *Environmentalism for a New Millennium. The Challenge of Coevolution*. Oxford University Press

Thirsk, J. 1997. *Alternative Agriculture. A History*. Oxford: Oxford University Press

Thomashow, M. 1995. *Ecological Identity*. Cambridge, MA: MIT Press

——. 2001. *Bringing the Biosphere Home. Learning to Perceive Global Environmental Change*. Cambridge, MA: MIT Press

Thu, K.M. and E.P. Durrenberger (eds). 1998. *Pigs, Profits, and Rural Communities*. Albany, N.Y.: State University of New York Press

Tilman, D. 1988. *Dynamics and Structure of Plant Communities*. Princeton: Princeton University Press

Tirosh-Samuelson, Hava (ed.) 2002. *Judaism and Ecology. Created World and Revealed World*. Cambridge, MA: Harvard University Press

Townsend, A.R.; R.W. Howarth, F.A. Bazzaz, M.S. Booth, C.C. Cleveland, S.K. Collinge, A.P. Dobson, P.R. Epstein, E.A. Holland, D.R. Keeney, M.A. Mallin, C.A. Rogers, P. Wayne, and A. Wolfe. 2003. "Human Health Effects of a Changing Global Nitrogen Cycle." *Frontiers in Ecology*, Issue 5, Volume 1 (June 2003): 240–246

Trevathan, Wenda, James J. McKenna and Euclid O. Smith (eds). 1999. *Evolutionary Medicine*. Oxford University Press

Trouillot, Michel-Rolph. 2003. *Global Transformations: Anthropology and the Modern World*. New York, Basingstoke: Palgrave Macmillan

Tucker, Mary Evelyn and John Berthrong (eds). 1998. *Confucianism and Ecology. The Interrelation of Heaven, Earth, and Humans*. Cambridge, MA: Harvard University Press

Tucker, Mary Evelyn and John A. Grim (eds). forthcoming. *Cosmology and Ecology*. Cambridge, MA: Harvard University Press

Tucker, Mary Evelyn and Duncan Ryuken Williams (eds). 1999. *Buddhism and Ecology. The Interconnection of Dharma and Deeds*. Cambridge, MA: Harvard University Press

Uhl, Christopher. 1988. "Restoration of Degraded Lands in the Amazon Basin." In: Wilson, E.O. (ed.) 1988. *Biodiversity*. Washington, D.C.: National Academy Press. Pp. 326–332

Ulrich, R.S. 1981. "Psychological and recreational benefits of a neighborhood park." *Journal of Leisure Research* 13: 43–65

——. 1984. "View through a window may influence recovery from surgery." *Science* 224: 420–421

Ulrich, R. et al. 1991. "Stress recovery during exposure to natural and urban environments." *Journal of Environmental Psychology* 11: 201–230

UNDP. 2000. *World Energy Assessment. Energy and the challenge of* sustainability.

UNEP (United Nations Environmental Programme), 2002. *GEO-3. Global Environment Outlook 3. Past, present, and future perspectives.* London and Sterling, VA: Earthscan Publications, Ltd.

Uphoff, N. 2003. "Higher Yields with Fewer External Inputs? The System of Rice Intensification and Potential Contributions to Agricultural Sustainability." *International Journal of Agricultural Sustainability*, Vol. 1 (1 November 2003), No. 1: 38–50

Van Boven, Leaf and Thomas Gilovich. 2003. "To Do or To Have? That Is the Question." *Journal of Personality and Social Psychology*, Vol. 85, No. 6 (2003): 1193–1202

Velimirov, Alberta and Werner Mueller. 2003. *Die Qualität biologisch erzeugter Lebensmittel. Ergebnisse einer umfassenden Literaturrecherche.* Vienna: Bio Ernte Austria [available on www.ernte.at]

Vickers, William T. 1994. "The Health Significance of Wild Plants for the Siona and Secoya." In: Etkin, Nina L. (ed.) *Eating on the Wild Side.* Tucson & London: The University of Arizona Press. Pp. 143–165

Vogel, Gretchen. 1998. "A Mini-Earth Struggles for Respectability" (Earth Science). *Science*, Vol. 280 (22 May 1998), No. 5367: 1183

Vogelsang, Keith M. "Footprint: Ignoring the Facts that don't fit the theory." *Nature* (Correspondence), Vol. 420 (21 November 2002): 267

Wackernagel, M., and W. Rees. 1996. *Our Ecological Footprint: Reducing Human Impact on the Earth.* Gabriola Island, B.C.: New Society Publishers

Wagner, Ulrich, Steffen Gais, Hilde Haider, Rolf Verleger and Jan Born. 2004. "Sleep inspires insight." *Nature*, Vol. 427 (22 January 2004): 352–355

Watts, Michael and Richard Peet. 1996. "Conclusion. Toward a theory of liberation ecology." In: Peet, Richard and Michael Watts (eds). *Liberation Ecologies.* London: Routledge. Pp. 260–269

Watzlawick, Paul. 1994 [1986]. Vom Schlechten des Guten oder Hekates Lösungen. Munich: dtv

WCED (World Commission on Environment and Development). 1987. *Our Common Future.* Oxford: Oxford University Press

Weiss, Harvey and Raymond S. Bradley. 2001. "What drives societal collapse?" *Science*, Vol. 291 (26 January 2001), No. 5504: 609–610

Wells, Nancy M. 2000. "At home with nature: effects of 'greenness' on children's cognitive functioning." *Environment and Behavior*, Vol. 32, No. 6 (November 2000): 775–795

Wells, Nancy M, and Gary W. Evans 2003. "Nearby Nature: A Buffer of Life Stress among Rural Children." *Environment and Behavior*, Vol. 35, No. 3 (1 May 2003): 311–330

Western, D. and M. Pearl (eds). 1989. *Conservation for the Twenty-first Century*. Oxford, New York: Oxford University Press

Whiteman, Gail. 1999. "Sustainability for the planet. A marketing perspective." *Conservation Ecology* 3(1): 13 [online] URL: www.consecol.org/vol3/iss1/art13

Wickler, Wolfgang and Uta Seibt. 1998. *Kalenderwurm und Perlenpost. Biologen entschlüsseln ungeschriebene Botschaften*. Heidelberg, Berlin: Spektrum Akademischer Verlag

Wiebe, G. 1973. "Mass media and Man's relationship to his environment." *Journalism Quarterly* 50: 426–432

Willis, K.J., L. Gillson and T.M. Brncic. "How 'Virgin' is Virgin Rainforest?" *Science*, Vol. 304 (16 April 2004): 402–403

Wilson, Edward O. (ed.) and Frances M. Peter (associate ed.) 1988. *Biodiversity* ("Papers from the National Forum on BioDiversity held September 21-25, 1986, in Washington, D.C., under the cosponsorship of the National Academy of Sciences and the Smithsonian Institution"). Washington, D.C.: National Academy Press

Wilson, Edward O. 1984. *Biophilia. The Human Bond With Other Species*. Cambridge, MA and London: Harvard University Press

——. 1998. *Consilience. The Unity of Knowledge*. New York: Alfred A. Knopf

——. 2002. *The Future of Life*. New York: Alfred A. Knopf

Wolfe, Martin S. 2000. "Crop strength through diversity." *Nature*, Vol. 406 (17 August 2000): 681–2

Wolff, Phillip and Douglas L. Medin. 2001. "Measuring the Evolution and Devolution of Folk-Biological Knowledge." In: Maffi, Luisa (ed). *On Biocultural Diversity. Linking language, knowledge, and the environment*. Washington and London: Smithsonian Institution Press. Pp. 212–227

World Commission on Dams. 2000. *Dams and Development. A New Framework for Decision-Making*. London: Earthscan

World Health Organization. 2001. *WHO Global Strategy for Containment of Antimicrobial Resistance*. [online] URL:
www.who.int/csr/drugresist/WHO_Global_Strategy_English.pdf

Woo, E. and M.J. Sharps. 2003. "Cognitive Aging and Physical Exercise." *Educational Gerontology*, Vol. 29, No. 4 (April 2003): 327–337

Wright, C. W (ed.) 2002. *Artemisia*. New York: Taylor & Francis

Yazdanbakhsh, Maria, Peter G. Kremsner and Ronald van Ree. 2002. "Allergy, Parasites and the Hygiene Hypothesis." *Science*, Vol. 296 (19 April 2002): 490–494

Young, T.K. 1993. "Diabetes mellitus among Native Americans in Canada and the United States. An epidemiological review." *American Journal of Human Biology* 5: 399–413

Zelezny, Lynnette C., and P. Wesley Schultz. 2000. "Promoting Environmentalism." *Journal of Social Issues*, Vol. 56, No. 3 (2000): 365–371

Zhu, Youyong; Hairu Chen, Jinghua Fan, Yunyue Wang, Yan Li, Jianbing Chen, JinXiang Fan, Shisheng Yang, Lingping Hu, Hei Leungk, Tom W. Mewk, Paul S. Tengk, Zonghua Wangk & Christopher C. Mundt. "Genetic diversity and disease control in rice." *Nature*, Vol. 406 (17 August 2000): 718–722

Index

Positive Ecology